Introduction to Toxicological Screening Methods and Good Laboratory Practice

Saravana Babu Chidambaram •
M. Mohamed Essa • M. Walid Qoronfleh

Introduction to Toxicological Screening Methods and Good Laboratory Practice

Springer

Saravana Babu Chidambaram
Department of Pharmacology,
JSS College of Pharmacy & Centre
for Experimental Pharmacology &
Toxicology
JSS Academy of Higher Education
and Research
Mysore, Karnataka, India

M. Mohamed Essa
College of Agricultural and Marine Sciences
Sultan Qaboos University
Al Khoudh, Muscat, Oman

M. Walid Qoronfleh
Q3CG Research Institute (QRI),
Research & Policy Division
Ann Arbor, MI, USA

ISBN 978-981-16-6091-7 ISBN 978-981-16-6092-4 (eBook)
https://doi.org/10.1007/978-981-16-6092-4

This Springer imprint is published by the registered company Springer Nature Singapore Pte Ltd.
The registered company address is: 152 Beach Road, #21-01/04 Gateway East, Singapore 189721, Singapore

Foreword

I was extremely delighted when I received a request from Dr. Saravana Babu Chidambaram to write a foreword for the book titled *Introduction to Toxicological Screening Methods and Good Laboratory Practice* which he had authored along with his collaborators, Dr. M. Mohammed Essa and Dr. M. Walid Qoronfleh.

I have been acquainted with Dr. Saravana Babu C for more than two decades, as my student, as a faculty of JSS Academy of Higher Education & Research, and as a fellow pharmacist. I have witnessed his scientific commitments, hard work and exemplary accomplishments in the fields of Neuropharmacology and Toxicology. Also, Dr. Babu has in-depth knowledge in establishing GLP standard preclinical toxicology facility and conducting toxicological studies as per the GLP principles and regulatory guidelines.

The Pharmacy Council of India has introduced Pharmacological and Toxicological Screening Methods and Toxicology as subjects for the postgraduate students and undergraduate students of Pharmacy, respectively. Although many books are currently available in the field of Toxicology, this book is unique in its contents that it provides the basic knowledge of toxicology, toxicological screening methods and Good Laboratory Practice elaborately yet in a simple language. The book focuses on the principles, screening methods and interpretation of data in establishing the risk and hazard assessment of small molecules as per the regulatory test guidelines, particularly OECD guidelines. Comprehensive information on toxicokinetics and safety pharmacology adds value to the book.

"Brevity in writing is the best insurance for its perusal"—Rudolf Virchow

I am confident the book will serve as a good reference material for students from disciplines like Medicine, Pharmacy and Life sciences interested in learning toxicology sciences. I appreciate the efforts of the authors to publish this book, which will be a great resource of comprehensive information on the Toxicology Screening Methods and Good Laboratory Practice.

Best wishes to the authors and the readers.

B. Suresh

JSS Academy of Higher Education & Research
Mysuru, Karnataka, India

Pharmacy Council of India
Mysuru, Karnataka, India

Foreword

It gives me immense pleasure to write a foreword for the book titled *Introduction to Toxicological Screening Methods and Good Laboratory Practice* authored by Drs. Saravana Babu Chidambaram, M. Mohamed Essa and M. Walid Qoronfleh. Toxicology testing is an indispensable component in the drug discovery process. Indeed, it is mandatory to establish the toxicology and safety pharmacology data through Phase I clinical trial, prior to the first exposure in humans. Also, systematic testing on the toxicological profile of a drug helps its market existence for a longer period of time which in turn requires strong foundation in the theoretical and practical aspects of toxicology sciences.

The book *Introduction to Toxicological Screening Methods and Good Laboratory Practice* is written in a comprehensive manner covering the basics of toxicology and its application in drug discovery, animal models used in toxicological screening and phenotypes, information on various regulatory bodies and their functions, short-term and long-term toxicological screening methods, genotoxicity, toxicokinetics, safety pharmacology and special toxicological models like reproductive and developmental toxicity, carcinogenicity and neurotoxicity studies. In addition, the book also provides basic yet elaborate information on Good Laboratory Practice and OECD principles of GLP.

In summary, the book *Introduction to Toxicological Screening Methods and Good Laboratory Practice* is a comprehensive scholarly effort by the leading scientists who have created a definitive resource material from their experience and expertise. I appreciate the authors for presenting the contents with great enthusiasm, particularly for the students in the disciplines of Medicine, Nursing, Pharmacy and Life sciences who wish to pursue their careers in the intriguing and fascinating field of toxicology.

I wish all the very best for the success of this wonderful endeavour and also the prospective readers.

JSS Academy of Higher Education & Research
Mysuru, Karnataka, India

Surinder Singh

Foreword

It is my pleasure to write a foreword for the book titled *Introduction to Toxicological Screening Methods and Good Laboratory Practice* authored by Dr. Saravana Babu C, Dr. M. Mohamed Essa and Dr. Walid Qoronfleh M. As a senior toxicologist, I have always understood the importance of and felt the need for a cohesive text that would impart the basics of toxicology to the undergraduate and postgraduate students. This book provides insights into the basics of non-clinical toxicology, animal models used in toxicology and their phenotypes, toxicology screening methods, toxicokinetics and safety pharmacology. The authors have also taken care to provide relevant and concise information about the Good Laboratory Practice. Each chapter of the book represents a particular topic which is written in simple language and has been well supported by figures, wherever required.

This book will be useful as a reference material for toxicological screening methods and Good Laboratory Practice for postgraduate and undergraduate students. I appreciate the hard work and goodwill of the authors in bringing out this book on toxicological screening methods.

Best wishes to the authors and the readers.

P. V. Mohanan

Toxicology Division, Sree Chitra Tirunal Institute for Medical Sciences & Technology, Biomedical Technology Wing
Thiruvananthapuram, Kerala, India

Dr. P. V. Mohanan
Scientist & Head, Toxicology Division
Sree Chitra Tirunal Institute for Medical
Sciences and Technology (Govt. of India),
Biomedical Technology Wing, Poojappura,
Thiruvananthapuram-695 012, Kerala, India

Preface

The tremendous growth in the healthcare system leading to increased human lifespan and long-term exposure to chemicals that are ubiquitous in the society prompt the use of precise, rapid, repeatable and cost-effective methods for potential risk and hazard assessment. Thus, the science of toxicology has emerged as an indispensable discipline in the drug discovery and development process. The science Toxicology helps to establish the safety window of potential hit molecules, thereby assisting in the identification of the lead compounds, using appropriate and validated in vitro and in vivo models

The book titled *Introduction to Toxicological Screening Methods and Good Laboratory Practice* focuses on the principles, methods and interpretations involved in establishing the safety, risk and hazard assessment of small molecules. The contents are presented with a thorough understanding on regulatory requirements for risk and hazard identification as per the Organization for Economic Co-operation and Development (OECD), Paris, International Council for Harmonisation (ICH) of Technical Requirements for Pharmaceuticals for Human Use ICH and Schedule 'Y", India, etc. guidelines.

The contents of this book will serve as reference material for the undergraduate and postgraduate pharmacy degree students to learn the principles, methods and interpretations of systemic toxicity (acute and repeated dose) and genotoxicity (in vitro and in vivo), special toxicological investigations such as reproductive and developmental toxicology, carcinogenicity, toxicokinetics using animal models or in vitro methods as applicable. This book also provides basic information on the basis of Good Laboratory Practice, OECD principles, conducting GLP toxicology study, and method of implementation of Good Laboratory Practice for academic understandings in a comprehensive manner.

This book contains selected definitions, abbreviated glossary, tables and figures/charts from OECD, ICH, etc. and test guidelines that are frequently used as toxicology reference materials. The content of the book does not attest the accuracy and/or completeness of such information and cannot assume any kind of liability arising on the use of such information. Details on the brands, vendors or any commercial products mentioned in the book do not reflect the authors' endorsement or recommendation for use. We hope this book could be a window for undergraduate and postgraduate medical, nursing, pharmacy and life sciences degree students for better understanding of the basics, terminologies and concepts in preclinical toxicology.

Mysuru, India — Saravana-Babu Chidambaram
Al Khoudh, Muscat, Oman — M. Mohamed Essa
Ann Arbor, MI, USA — M. Walid Qoronfleh

Acknowledgement

First and foremost, we would like to praise and thank God, the Almighty, who has blessed us with the knowledge and opportunity to come together and write this book.

We as authors thankfully want to acknowledge our respective institutions for their continued support that allowed us to finish this book in a successful and timely manner.

Many thanks are due to our mentors who have contributed at various junctures in our careers and have inspired us to be curious to dream big and to explore more.

Our heartfelt gratitude to our colleagues with whom we have had many stimulating discussions and enriching conversations.

We trust the contents of this book will serve as reference material for the undergraduate and postgraduate students of Medicine, Nursing, Pharmacy and Life sciences for better understanding of the basics, terminologies and concepts pertaining to preclinical toxicology screening methods and Good Laboratory Practices and wish the prospective readers a very good luck in the all endeavours.

Our heartfelt thanks to The Editing Refinery, MD, USA for providing meticulous assistance in proof reading and language editing of the text book.

We have utilized BioRender software for creating a few figures and we immensely thank BioRender.com for the same.

We sincerely thank the staff of Springer Nature, Singapore, for their patience and assistance at various stages of the book publication.

We are highly indebted to our families for their understanding and unconditional support during this endeavour allowing us to spend extra time on completing this book and fulfil our dream.

Contents

Introduction to Toxicology

1

1.1 Definition

Toxicology (In Greek: Toxicon—poison; Logos—study) is an interdisciplinary field of biomedical sciences that is concerned with the study of adverse effects of chemicals (pharmaceuticals, industrial chemicals, pesticides, food additives, etc.), biologicals (toxins e.g. poisonous plants and venomous products of animal origin) or physical agents (e.g. radiation, noise, etc.) on living systems upon direct or indirect exposure (Gowda et al. 2014; Roberts et al. 2014). The adverse effects might range from immediate death to subtle changes that are unrealised until months or years later.

Adverse effects of a test item depends on two main aspects:

- Routes of administration (inhalation, oral, dermal or parenteral) and
- Dose (concentration of administration and duration)

In order to understand the dose-dependent effects, the test items are generally tested in both acute and chronic treatment models. In addition, other factors like species, strain, sex, age, body weight, environment, chronobiology, individual physiological characters and environment also influence the adverse outcomes of a test item. Depending on the specific target organ and endpoints, there are various special toxicological studies viz. carcinogenicity, genotoxicity, neurotoxicity, nephrotoxicity, immunotoxicity, reproductive and developmental toxicity.

Depending on the duration and number of doses administered, toxicology studies are divided into two major types (Table 1.1):

1. Acute (one-time dosing; minimum 14 days experimentation in rodents) and (Frank and Ottoboni 2011; Hodgson 2010)
2. Repeated dose (multiple dosing; 7 days to 2 years in rodents) toxicity studies

S. B. Chidambaram et al., *Introduction to Toxicological Screening Methods and Good Laboratory Practice*, https://doi.org/10.1007/978-981-16-6092-4_1

Table 1.1 Types of toxicology studies

Acute toxicology	Repeated dose Classical toxicology	Special toxicology
• Acute—Oral • Acute—Dermal toxicity • Acute—Dermal irritation/corrosion • Acute inhalation • Skin sensitisation • Eye irritation/ corrosion	• 14/28-day repeated dose (sub-acute toxicity) • 90-day repeated dose (subchronic toxicity) • 180 and above—repeated dose (chronic toxicity)	• Genotoxicity • Carcinogenicity • Neurotoxicity • Reproductive and developmental toxicity • Immunotoxicity

1.2 History

The historical evidence of toxicology began early with Dioscorides, a Greek physician, and was the first to categorise poisons with descriptions and drawings (Levey 1966). Dioscorides' classification of plants, animals and mineral poisons were not only left as standard for 16 centuries but also have stood the test of time and is still a convenient classification (James et al. 2000). Paracelsus, a Swiss Physician–Alchemist (1493–1541) (Borzelleca 2000) popularly referred to as 'Father of Toxicology' established a number of new perspectives that are still part of the integral structure of modern toxicology, pharmacology and therapeutics (Pagel 1958 n.d.). 'Paracelsus' is often quoted for his statement: 'All substances are poisons; there is none which is not a poison. The right dose differentiates a poison and a remedy' (Professionals et al. 1997). Mathieu Orfila, a Spanish physician, known as the 'Father of Modern Toxicology' systematically described the relationship between chemical and biological effects of toxicants with respect to specific organ related toxicity by analysing autopsy samples (Wennig 2009). Since the nineteenth and twentieth centuries, many German scientists have contributed immensely to the growth of toxicological sciences. Oswald Schmiedeberg (1838–1921) and Louis Lewin (1850–1929) were more popular toxicologists during this period. The basic knowledge in understanding the effects of toxicants at cellular and molecular levels were developed in the late twentieth century. The thalidomide tragedy of birth defects published as 'Silent Spring' by Rachel Carson created alarming awareness on the importance of toxicology in the areas of drug discovery and development. Attempts to understand the effects of chemicals on the reproductive and developmental biology and on the ecology also gained major attention. Establishment of various regulatory bodies like Environmental Protection Agency (EPA), U.S; International Conference on Harmonisation (ICH), U.S; Food and Drug Administration, National Toxicology Program (NTP), Organization for Economic Cooperation and Development (OECD), etc. added scientific and legal strength to toxicological sciences.

1.3 Types of Toxicology (Barile 2019)

With the development of modern biotechnology, precise, and ultrafast analytical techniques, toxicology information became highly important in various fraternities like drug discovery, forensic medicine, land and marine ecology, including legal aspects, etc. Based on the applications and endpoints, more accurately labelled sub-fields in toxicological sciences emerged viz., regulatory toxicology, mechanistic toxicology, descriptive toxicology, forensic toxicology and clinical toxicology.

1.3.1 Descriptive Toxicology

All systemic and environmental toxicological investigations that provide details on the potential risk and health hazard to a living system by the administration of a test item are covered under descriptive toxicology. It has other sub-descriptive fields viz. genetic toxicology, carcinogenicity, immunotoxicity, neurotoxicity, nephrotoxicity, hepatotoxicity, reproductive and developmental toxicity. Descriptive toxicology deploys clinical biochemistry, pathology, genes and proteins expression techniques to provide the potential risk and hazard information of a test item (Roberts et al. 2014).

1.3.2 Mechanistic Toxicology

This branch of toxicology science studies the mechanism of action of drugs/poisons and how it adversely affects cellular, physiological and behavioural functions.

1.3.3 Clinical Toxicology

It deals with overdosing of drugs and poisons, identification of toxicants in the biological matrices, information to handle the emergencies by studying the signs and symptoms caused by toxicants/toxins and remedial information, at least partly.

1.3.4 Forensic Toxicology

This branch of toxicological science deals with the identification and estimation of toxicants in the biological matrices (autopsy specimens like—hair, urine, blood, bone or organ; tissue samples that are collected, maintained and disposed under the supervision of medical experts/pathologists). Usually, Board-certified forensic chemists send testing samples to a toxicology laboratory.

The analytical investigations are performed using sophisticated analytical techniques and with legally defensible standards, including initial and confirmatory testing. Forensic toxicology utilises techniques like Gas chromatography (GC), Gas chromatography and mass spectrometry (GC-MS/MS), enzyme-linked immunosorbent assay (ELISA), high-performance thin layer chromatography (HPLC), liquid chromatography and mass spectrometry (LC-MS/MS), etc. for the bio-analytical procedures.

1.3.5 Regulatory Toxicology

This branch of toxicological sciences provides information on the potential risk and hazard of a test item (including pharmaceuticals, food additives, industrial chemicals, pesticides and cosmetics) likely to occur to the living system and environment up on acute or repeated exposure. The regulatory toxicological studies are performed following definite test guidelines established by the federal/government authorities. For the estimation of the risk potential of a test item, information such as reversible and irreversible effects, low-observed adverse effect level (LOAEL), no-observed effect level (NOAEL), margin of exposure (MoE), margin of safety (MoS), etc. are developed and the data are utilised for regulatory submissions. The regulatory dossier submission, with respect to preclinical toxicological data, varies between countries or regulatory bodies so as to fulfil the respective testing guidelines (Barile 2019).

1.4 Standard, Test Guidelines and Regulatory Bodies

1.4.1 Standards

The Standards provide required specifications and quantifiable measures in toxicological testing. Organisations that set standards for toxicology testing are generally regional specific, government, quasi-government, non-government bodies or combinations of the bodies (Table 1.2).

1.4.2 Test Guidelines (TG)

With respect to the toxicology domain, the organisations (generally referred to as regulatory bodies) that develop guidelines for toxicology testing are global, regional

Table 1.2 Standard setting organisations [Adapted from (Pleus 2014)]

Location	Names of standard setting organization	Web link
Global	ISO	https://www.iso.org/home.html
	ASTM International	https://www.astm.org/
	CLSI	http://www.clsi.org/
Country	BSI	http://www.bsigroup.com/
	CSA	http://www.csa.ca/cm/ca/en/home

Table 1.3 Toxicity test guidelines setting organisations [Adapted from (Pleus 2014)]

Location	Names of test guideline setting organization	References
Global	ICH	https://www.ich.org/
	OECD	https://www.oecd.org/home/
	WHO	http://www.who.int
Regional country	EU IHCP	http://ihcp.jrc.ec.europa.eu/home
	US EPA OCSPP	http://www.epa.gov/aboutepa/ocspp.html
	US FDA	http://www.fda.gov/
	NTP	http://ntp.niehs.nih.gov/

and country-specific (Table 1.3). Test guidelines provide details on the methodology, performance and monitoring procedures to be followed during toxicological testing.

TGs are developed by experts based on robust and rigorous methods, validated procedures and evidence-based approaches. The draft guidelines are posted for public review and reviewed by independent experts. Then, the appropriate review comments/suggestions are incorporated, published and distributed. In general, toxicological test guidelines provide clear information on preferred species, numbers of animals per group, housing, acclimatisation and randomisation, groups and dosing, observations, statistical and reporting procedures. TGs are periodically (5–10 years) revised and updated.

The important test guidelines related to human and environment safety–related toxicity tests developed by various regulatory bodies are presented in Table 1.4.

1.5 Organisation for Economic Co-operation and Development (OECD)

OECD is an intergovernmental economic organisation, officially established in 1961 with 18 countries representing from United States, Europe, and Canada (founding members) on the forum to create an organisation dedicated to economic development. The headquarters of OECD is located in Paris, France, and the regional centres are in Berlin, Mexico, Tokyo and Washington.

Official founding members:

Austria	West Germany	Luxembourg	Sweden
Belgium	Greece	The Netherlands	Switzerland
Canada	Iceland	Norway	Turkey
Denmark	Ireland	Portugal	United Kingdom
France	Italy	Spain	United States

Membership countries:

As of March 2021, from North and South America to Europe and Asia-Pacific 37 member countries are spanning across the globe. It includes many developed countries and also developing countries like Mexico, Chile and Turkey.

Austria	Estonia	Israel	Netherlands	Sweden
Australia	Finland	Italy	New Zealand	Switzerland
Belgium	France	Japan	Norway	Turkey
Canada	Germany	Korea	Poland	United Kingdom
Chile	Greece	Latvia	Portugal	United States
Colombia	Hungary	Lithuania	Slovak Republic	
Czech Republic	Iceland	Luxembourg	Slovenia	
Denmark	Ireland	Mexico	Spain	

The OECD member countries committed to democracy and the social economy by offering a platform to share policy experiences, discover solutions to challenges, recognise good laboratory practices and coordinate their domestic and international policies.

1.5.1 Organisational Structure

It functions on a three-level structure, viz. the Council, the Secretariat and the Committees (Fig. 1.1).

1.5.2 The Council

It comprises ambassadors from member countries, who are involved in establishing the goals of OECD, strategic planning and decision making.

1.5.3 The Secretariat

It comprises Secretary—General, the deputy and the directorates. The Secretariat includes 2500 plus members, which include scientists, lawyers, economists involved in research conduction, monitoring and data analysis. The General Secretary oversees the functions of the secretariat.

1.5.4 The Committees

It comprises representatives from member countries, which function to discuss the environment, education, trade and investment.

Table 1.4 Important toxicity test guidelines related to human and environment safety [Adapted from (Pleus 2014)]

Toxicological test	Organizations that have guidelines	Web address
Acute	US EPA OCSPP	http://www.regulations.gov/#!documentDetail; D¼EPA-HQ-OPPT-2009-01560002
	OECD	http://www.oecd-ilibrary.org/environment/test-no-403-acute-inhalation
	NTP	http://ntp.niehs.nih.gov/?objectid¼363107BD-F1F6-975E-7EEFC3A4FD87ADA2
	US FDA	http://www.fda.gov/Food/GuidanceComplianceRegulatoryInformation/GuidanceDocuments/FoodIngredientsandPackaging/Redbook/ucm078339.htm
Subacute	US EPA OCSPP	http://www.regulations.gov/#!documentDetail; D¼EPA-HQ-OPPT-2009-0156-0012
	OECD	http://www.oecd-ilibrary.org/environment/test-no-412-subacute-inhalation-toxicity-28-daystudy_9789264070783-en
	OECD	http://www.oecd-ilibrary.org/environment/test-no-413-subchronic-inhalation-toxicity-90-day-study_9789264070806-en
	NTA	http://ntp.niehs.nih.gov/?objectid¼3630B364-F1F6-975E-7CD06271314075A2
	FDA	http://www.fda.gov/Food/GuidanceComplianceRegulatoryInformation/GuidanceDocuments/FoodIngredientsandPackaging/Redbook/ucm078345.htm
Chronic	US EPA OCSPP	http://www.regulations.gov/#!documentDetail; D¼EPA-HQ-OPPT-2009-0156-0019
	OECD	http://www.oecd-ilibrary.org/environment/test-no-452-chronic-toxicity-studies_9789264071209-en
	ICH	http://www.ich.org/products/guidelines/safety/article/safety-guidelines.html
	NTP	http://ntp.niehs.nih.gov/?objectid¼36305D16-F1F6-975E-79776DAD38EC101E
	FDA	http://www.fda.gov/Food/GuidanceComplianceRegulatoryInformation/GuidanceDocuments/FoodIngredientsandPackaging/Redbook/ucm078349.htm
Neurotoxicity	US EPA OCSPP	http://www.regulations.gov/#!documentDetail; D¼EPA-HQ-OPPT-2009-0156-0041
	OECD	http://www.oecd-ilibrary.org/environment/test-no-424-neurotoxicity-study-in-rodents_9789264071025-en
	FDA	http://www.fda.gov/Food/GuidanceComplianceRegulatoryInformation/GuidanceDocuments/FoodIngredientsandPackaging/Redbook/ucm078323.htm

(continued)

Table 1.4 (continued)

Toxicological test	Organizations that have guidelines	Web address
	WHO	http://www.inchem.org/documents/ehc/ehc/ehc060.htm
Reproductive	WHO	http://www.inchem.org/documents/ehc/ehc/ehc060.htm
	US EPA OCSPP	http://www.regulations.gov/#!documentDetail;D¼EPA-HQ-OPPT-2009-0156-0042
	OECD	http://www.oecd-ilibrary.org/environment/test-no-414-prenatal-development-toxicity-study_9789264070820-en
	ICH	http://www.ich.org/products/guidelines/safety/article/safety-guidelines.html
	NTP	http://ntp.niehs.nih.gov/?objectid¼070C9F1C-A6F9-B112-990A9F55BAC7C407
	FDA	http://www.fda.gov/Food/GuidanceComplianceRegulatoryInformation/GuidanceDocuments/
	WHO	http://www.inchem.org/documents/ehc/ehc/ehc30.htm
Immunotoxicity	US EPA OCSPP	http://www.regulations.gov/#!documentDetail;D¼EPA-HQ-OPPT-2009-0156-0049
	ICH	http://www.ich.org/products/guidelines/safety/article/safety-guidelines.html
	WHO	http://www.inchem.org/documents/ehc/ehc/ehc180.htm
Genotoxicity	US EPA OCSPP	http://www.regulations.gov/#!documentDetail;D¼EPA-HQ-OPPT-2009-0156-0028
	OECD	http://www.oecd-ilibrary.org/environment/test-no-488-transgenic-rodent-somatic-and-germ-cellgene-mutation-assays_9789264122819-en
	ICH	http://www.ich.org/products/guidelines/safety/article/safety-guidelines.html
Inhalation	US EPA OCSPP	http://www.regulations.gov/#!documentDetail;D¼EPA-HQ-OPPT-2009-0156-0014
	OECD	http://www.oecd-ilibrary.org/environment/test-no-413-subchronic-inhalation-toxicity-90-day-study_9789264070806-en
Dermal	US EPA OCSPP	http://www.regulations.gov/#!documentDetail;D¼EPA-HQ-OPPT-2009-0156-0004
	OECD	http://www.oecd-ilibrary.org/environment/test-no-402-acute-dermaltoxicity_9789264070585-en
Oral	US EPA OCSPP	http://www.regulations.gov/#!documentDetail;D¼EPA-HQ-OPPT-2009-0156-0010
	OECD	http://www.oecd-ilibrary.org/environment/test-no-408-repeated-dose-90-day-oral-toxicity-study-inrodents_9789264070707-en
Carcinogenicity	US EPA OCSPP	http://www.regulations.gov/#!documentDetail;D¼EPA-HQ-OPPT-2009-0156-0020

(continued)

Table 1.4 (continued)

Toxicological test	Organizations that have guidelines	Web address
	OECD	http://www.oecd-ilibrary.org/environment/test-no-451-carcinogenicitystudies_9789264071186-en
	ICH	http://www.ich.org/products/guidelines/safety/article/safety-guidelines.html
	US FDA	http://www.fda.gov/Food/GuidanceComplianceRegulatoryInformation/GuidanceDocuments/FoodIngredientsandPackaging/Redbook/ucm078388.htm
In vitro	US EPA OCSPP	http://www.regulations.gov/#!documentDetail;D¼EPA-HQ-OPPT-2009-0156-0038
	OECD	http://www.oecd-ilibrary.org/environment/test-no-476-in-vitro-mammalian-cell-gene-mutation-test_9789264071322-en

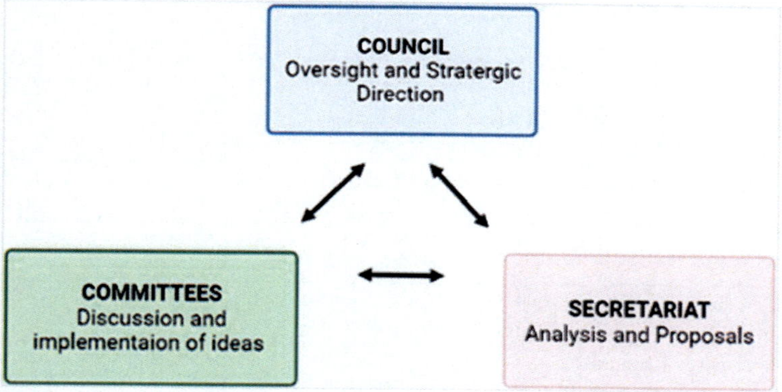

Fig. 1.1 OECD organisational structures. [Adapted from (OECD - Overview n.d.)]

1.5.5 Functions of OECD

- Developing guidelines from the most relevant and widely accepted international testing methodologies by governments, industry and independent laboratories for testing of chemicals. The guidelines are utilised in regulatory studies and following chemical notification and registration.
- Assisting member countries to perform the validated testing methods in their laboratories, which ensure high quality and reliable data for the safety of chemicals.

Table 1.5 Section 1—Physical–Chemical Properties [Adapted from (OECD Guidelines for the Testing of Chemicals, Section 1 n.d., p. 1)]

No.	Title	Original adoption	No. of updates	Updates	Most recent update
101	UV-VIS Absorption Spectra	12 May 1981	0		–
102	Melting Point/Melting Range	12 May 1981	1		27 July 1995
103	Boiling Point	12 May 1981	1		27 July 1995
104	Vapour Pressure	12 May 1981	2	27 July 1995	23 March 2006
105	Water Solubility	12 May 1981	1		27 July 1995
106	Adsorption/Desorption Using a Batch Equilibrium Method	12 May 1981	1		21 January 2000
107	Partition Coefficient (n-octanol/ water): Shake Flask Method	12 May 1981	1		27 July 1995
108	Complex Formation Ability in Water	12 May 1981	0		–
109	Density of Liquids and Solids	12 May 1981	2	27 July 1995	2 October 2012
110	Particle Size Distribution/Fibre Length and Diameter Distributions	12 May 1981	0		–
111	Hydrolysis as a function of pH	12 May 1981	1		13 April 2004
112	Dissociation Constants in Water	12 May 1981	0		–
113	Screening Test for Thermal Stability and Stability in Air	12 May 1981	0		–
114	Viscosity of Liquids	12 May 1981	1		2 October 2012
115	Surface Tension of Aqueous Solutions	12 May 1981	1		27 July 1995
116	Fat Solubility of Solid and Liquid Substances	12 May 1981	0		–
117	Partition Coefficient (n- octanol/ water), HPLC method	30 March 1989	1		13 April 2004
118	Determination of the Number-Average Molecular Weight and the Molecular Weight Distribution of Polymers using Gel Permeation Chromatography	14 June 1996	0		–
119	Determination of the Low Molecular Weight Content of a Polymer Using Gel Permeation Chromatography	14 June 1996	0		–

(continued)

Table 1.5 (continued)

No.	Title	Original adoption	No. of updates	Updates	Most recent update
120	Solution/Extraction Behaviours of Polymers in Water	14 June 1996	1		21 January 2000
121	Estimation of the Adsorption Coefficient (K_{OC}) on Soil and on Sewage Sludge using High-Performance Liquid Chromatography (HPLC)	22 January 2001	0		–
122	Determination of pH, Acidity and Alkalinity	26 July 2013	0		–
123	Partition Coefficient (1-Octanol/ Water): Slow-Stirring Method	23 March 2006	0		–

- OECD continuously reviews and revises the methodologies to ensure more safe data.
- Approval of Mutual Acceptance of Data (MAD) to avoid duplication of studies, which allows industries and countries to fully benefit from the economic point of view. MAD also helps to minimise the number of animals used for the testing.
- Promoting local and regional development
- Combating international tax avoidance
- Guiding economic reforms of nations

The OECD Guidelines for the Testing of Chemicals ("OECD Test Guidelines for Chemicals - OECD," n.d.) have been classified under five sections (Tables 1.5, 1.6, 1.7, 1.8, and 1.9)

1. Section 1: Physical–Chemical Properties (OECD Guidelines for the Testing of Chemicals, Section 1 n.d., p. 1)
2. Section 2: Effects on Biotic Systems (OECD Guidelines for the Testing of Chemicals, Section 1 n.d., p. 2)
3. Section 3: Environmental fate and behaviour (OECD Guidelines for the Testing of Chemicals, Section 3 n.d., p. 3)
4. Section 4: Health effects (OECD Guidelines for the Testing of Chemicals, Section 4 n.d., p. 4)
5. Section 5: Other Test Guidelines (OECD Guidelines for the Testing of Chemicals, Section 5 n.d., p. 5)

1.5.6 OECD Guidelines for Testing for Chemicals

Table 1.6 Section 2—Effects on Biotic Systems [Adapted from (OECD Guidelines for the Testing of Chemicals, Section 2 n.d., p. 2)]

No.	Title	Original adoption	No. of updates	Updates	Most recently updated
201	Freshwater Alga and Cyanobacteria, Growth Inhibition Test	12 May 1981	2	7 June 1984	23 March 2006 (corrected on 28 July 2011)
202	*Daphnia* sp. Acute Immobilisation Test	12 May 1981 (adopted as *Daphnia* sp.14-day Reproduction Test including an Acute Immobilisation Test)	2	4 April 1984	13 April 2004
203	Fish, Acute Toxicity Test	12 May 1981	2	4 April 1984	17 July 1992
204	Fish, Prolonged Toxicity Test: 14-Day Study	4 April 1984	0		Date of deletion 2 April 2014
205	Avian Dietary Toxicity Test	4 April 1984	0		–
206	Avian Reproduction Test	4 April 1984	0		–
207	Earthworm, Acute Toxicity Tests	4 April 1984	0		–
208	Terrestrial Plants, Growth Test	4 April 1984	1		19 July 2006
209	Activated Sludge, Respiration Inhibition Test (Carbon and Ammonium Oxidation)	4 April 1984	1		22 July 2010
210	Fish, Early-Life Stage Toxicity Test	17 July 1992	1		26 July 2013
211	*Daphnia magna* Reproduction Test	21 September 1998	2	3 October 2008	2 October 2012
212	Fish, Short-term Toxicity Test on Embryo and Sac-fry Stages	21 September 1998	0		–
213	Honeybees, Acute Oral Toxicity Test	21 September 1998	0		–
214	Honeybees, Acute Contact Toxicity Test	21 September 1998	0		–
215	Fish, Juvenile Growth Test	21 January 2000	0		–

(continued)

Table 1.6 (continued)

No.	Title	Original adoption	No. of updates	Updates	Most recently updated
216	Soil Microorganisms: Nitrogen Transformation Test	21 January 2000	0		–
217	Soil Microorganisms: Carbon Transformation Test	21 January 2000	0		–
218	Sediment-Water Chironomid Toxicity Using Spiked Sediment	13 April 2004	0		–
219	Sediment-Water Chironomid Toxicity Using Spiked Water	13 April 2004	0		–
220	Enchytraeid Reproduction Test	13 April 2004	1		26 July 2016
221	*Lemna* sp. Growth Inhibition Test	23 March 2006	0		–
222	Earthworm Reproduction Test (*Eiseniafetida/ Eiseniaandrei*)	13 April 2004	1		26 July 2016
223	Avian Acute Oral Toxicity Test	22 July 2010	1		26 July 2016
224	Determination of the activity of anaerobic bacteria—Reduction of gas production from anaerobically sewage sludge	8 January 2007	0		–
225	Sediment-Water *Lumbriculus* Toxicity Test Using Spiked Sediment	16 October 2007	0		–
226	Predatory Mite (*Hypoaspis (Geolaelaps) Aculeifer*) Reproduction Test in Soil	3 October 2008	1		26 July 2016
227	Terrestrial Plant Test: Vegetative Vigour Test	19 July 2006	0		–
228	Determination of Developmental Toxicity of a Test Chemical to Dipteran Dung Flies (*Scathophagastercoraria* L. (*Scathophagidae*) and *Musca automnalis* De Geer (*Muscidae*))	3 October 2008	1		26 July 2016
229	Fish Short-Term Reproduction Assay	7 September 2009	1		2 October 2012

(continued)

Table 1.6 (continued)

No.	Title	Original adoption	No. of updates	Updates	Most recently updated
230	21-Day Fish Assay: A Short-Term Screening Assay for Oestrogenic and Androgenic Activity, and Aromatase Inhibition	7 September 2009	0		–
231	Amphibian Metamorphosis Assay	7 September 2009	0		–
232	Collembolan Reproduction Test in Soil	7 September 2009	1		26 July 2016
233	Sediment-Water Chironomid Life-Cycle Toxicity Test Using Spiked Water or Spiked Sediment	22 July 2010	0		–
234	Fish Sexual Development Test	28 July 2011	0		–
235	*Chironomus* sp., Acute Immobilisation Test	28 July 2011	0		–
236	Fish Embryo Acute Toxicity (FET) Test	26 July 2013	0		–
237	Honey bee (*Apismellifera*) Larval Toxicity Test, Single Exposure	26 July 2013	0		–
238	Sediment-free *Myriophyllum spicatum* Toxicity Test	26 September 2014	0		–
239	Water-Sediment *Myriophyllum spicatum* Toxicity Test	26 September 2014	0		–
240	Medaka Extended One-Generation Reproduction Test	28 July 2015	0		–
241	Larval Amphibian Growth and Development Assay	28 July 2015	0		–
242	*Potamopyrgu santipodarum* Reproduction Test	26 July 2016	0		–
243	*Lymnaea stagnalis* Reproduction Test	26 July 2016	0		–
244	Protozoa Activated Sludge Inhibition Test	9 October 2017	0		–
245	Honeybee, 10-day oral toxicity test	9 October 2017	0		–
246	Bumblebee, acute contact toxicity test	9 October 2017	0		–

(continued)

Table 1.6 (continued)

No.	Title	Original adoption	No. of updates	Updates	Most recently updated
247	Bumblebee, acute oral toxicity test	9 October 2017	0		–

1.6 International Conference on Harmonisation (ICH) (ICH Guidelines for Pharmaceuticals: Pharmaceutical Guidelines n.d.)

International Council for Harmonisation (ICH) was formerly known as International Conference on Harmonisation (ICH) of technical requirements for registration of pharmaceuticals for human use. It is a joint initiative of both regulators and representatives from research-based industries in the European Union, Japan and United States, to conduct conferences to discuss scientific and technical aspects of the test methods needed to examine and ensure the safety, quality and efficacy of medications.

ICH constitutes representatives from six co-sponsoring parties (**EU**: European Commission [EC] and European Federation of Pharmaceutical Industries' Associations [EFPIA]; **Japan**: Ministry of Health & Welfare [MHW] and Japan Pharmaceutical Manufacturers Association [JPMA], three observers and the International Federation of Pharmaceutical Manufacturers Associations (IFPMA), **USA**: Food and Drug Administration [FDA] and Pharmaceutical Research and Manufacturers of America [PhRMA], and Observers from World Health Organization (WHO), European Free Trade Association (EFTA), and Canada. The Observers represent the non-ICH countries and regions.

1.6.1 History

Europe, Japan and the United States proposed to develop harmonisation plans in 1989, which resulted in the birth of the International Council of Harmonisation of Technical Requirements for Registration of Pharmaceuticals for Human Use (ICH), in April 1990 in Brussels. Since the inception, ICH has made consistent progress. In the first decade, ICH has been made significant progress in the development of Safety, Quality and Efficacy guidelines. Several other important multidisciplinary topics, such as the Medical Dictionary for Regulatory Activities, simultaneously (MedDRA) and Common Technical Document (CTD) were developed. In the second decade, ICH concentrated on facilitating the implementation of ICH Guidelines in its own regions. ICH intends to enhance communication and dissemination of ICH Guidelines information to non-ICH regions in the fourth decade. In 2015, a significant step has been taken to facilitate a series of organisational changes

Table 1.7 Section 3—Environmental Fate and Behaviour [Adapted from (OECD Guidelines for the Testing of Chemicals, Section 3 n.d., p. 3)]

No.	Title	Original adoption	No. of updates	Updates	Most recently updated
301	Ready Biodegradability: 301A: DOC Die-Away Test 301B: CO_2 Evolution Test 301C: Modified MITI Test (I) 301D: Closed Bottle Test 301E: Modified OECD Screening Test 301F: Manometric Respirometry Test	12 May 1981	1		17 July 1992 (301F corrected on 26 July 2013)
302A	Inherent Biodegradability: Modified SCAS Test	12 May 1981	0		–
302B	Inherent Biodegradability: Zahn-Wellens/EMPA Test	12 May 1981	1		17 July 1992
302C	Inherent Biodegradability: Modified MITI Test (II)	12 May 1981	0		(corrected in 2009)
303	Simulation Test—Aerobic Sewage Treatment: A: Activated Sludge Units; B: Biofilms	12 May 1981	1		22 January 2001
304A	Inherent Biodegradability in Soil	12 May 1981	0		–
305	Bioaccumulation in Fish: Aqueous and Dietary Exposure	12 May 1981	2	14 June 1996	2 October 2012
306	Biodegradability in Seawater	17 July 1992	0		–
307	Aerobic and Anaerobic Transformation in Soil	24 April 2002	0		–
308	Aerobic and Anaerobic Transformation in Aquatic Sediment Systems	24 April 2002	0		–
309	Aerobic Mineralisation in Surface Water—Simulation Biodegradation Test	13 April 2004	0		
310	Ready Biodegradability—CO2 in sealed vessels (Headspace Test)	23 March 2006	0		(Corrected on 26 September 2014)
311	Anaerobic Biodegradability of Organic Compounds in Digested Sludge: by Measurement of Gas Production	23 March 2006	0		–
312	Leaching in Soil Columns	13 April 2004	0		–
313	Estimation of Emissions from Preservative—Treated Wood	16 October 2007	0		–

(continued)

Table 1.7 (continued)

No.	Title	Original adoption	No. of updates	Updates	Most recently updated
	to the Environment: Laboratory Method for Wooden Commodities that are not Covered and are in Contact with Fresh Water or Seawater				
314	Simulation Tests to Assess the Biodegradability of Chemicals Discharged in Wastewater	3 October 2008	0		–
315	Bioaccumulation in Sediment-dwelling Benthic Oligochaetes	3 October 2008	0		–
316	Photo transformation of Chemicals in Water- Direct Photolysis	3 October 2008	0		–
317	Bioaccumulation in Terrestrial Oligochaetes	22 July 2010	0		–
318	Dispersion Stability of Nanomaterials in Simulated Environmental Media	9 October 2017	0		–
319A	Determination of in vitro intrinsic clearance using cryopreserved rainbow trout hepatocytes (RT-HEP)	27 June 2018	0		–
319B	Determination of in vitro intrinsic clearance using rainbow trout liver S9 sub-cellular fraction (RT-S9)	27 June 2018	0		–

that include expansion of international outreach, revamping ICH's governance structure, circulating ICH processes to a broader range of stakeholders and establishing ICH as a legal entity for a more stable operating structure.

1.6.2 ICH Organisational Structure and Functions

The organisational structure of ICH consists of ICH Steering Committee, ICH Coordinators, ICH Secretariat and ICH Working Groups (Fig. 1.2)

1.6.2.1 ICH Steering Committee

The Steering Committee constitutes two representatives from each co-sponsoring party. It governs the functions of ICH, establishes the ICH plan and procedures, determines the harmonisation topics and analyses the implementation of the harmonisation efforts. The Steering Committee meets at least twice a year.

Table 1.8 Section 4—Health Effects [Adapted from (OECD Guidelines for the Testing of Chemicals, Section 4 n.d., p. 4)]

No.	Title	Original adoption	No. of updates	Updates	Most recently updated
401	Acute Oral Toxicity	12 May 1981	1	24 February 1987	Date of Deletion: 17 December 2001
402	Acute Dermal Toxicity	12 May 1981	2	24 February 1987	9 October 2017
403	Acute Inhalation Toxicity	12 May 1981	1		7 September 2009
404	Acute Dermal Irritation/Corrosion	12 May 1981	3	17 July 1992 24 April 2002	28 July 2015
405	Acute Eye Irritation/ Corrosion	12 May 1981 (corrected in 2017)	3	24 April 2002 24 Feb. 1987	2 Oct. 2012
406	Skin Sensitisation	12 May 1981	1		17 July 1992
407	Repeated Dose 28-Day Oral Toxicity Study in Rodents	12 May 1981	2	27 July 1995	3 October 2008
408	Repeated Dose 90-Day Oral Toxicity Study in Rodents	12 May 1981	2	21 September 1998	25June 2018
409	Repeated Dose 90-Day Oral Toxicity Study in Non-Rodents	12 May 1981	1		21 September 1998
410	Repeated Dose Dermal Toxicity:90-Day	12 May 1981	0		–
411	Subchronic Dermal Toxicity: 90-day Study	12 May 1981	0		–
412	Subacute Inhalation Toxicity: 28-Day Study	12 May 1981	2	7 September 2009	9 October 2017 (Corrected 27 June 2018)
413	Subchronic Inhalation Toxicity: 90-Day Study	12 May 1981	2	7 September 2009	9 October 2017 (Corrected 27 June 2018)
414	Prenatal Developmental Toxicity Study	12 May 1981	2	22 January 2001	27 June 2018
415	One-Generation Reproduction Toxicity Study	26 May 1983	0		Date of Deletion: 9 October 2017

(continued)

Table 1.8 (continued)

No.	Title	Original adoption	No. of updates	Updates	Most recently updated
					(effective 9 April 2019)
416	Two-Generation Reproduction Toxicity Study	26 May 1983	1		22 January 2001
417	Toxicokinetics	4 April 1984	1		22 July 2010
418	Delayed Neurotoxicity of Organophosphorus Substances Following Acute Exposure	4 April 1984	1		27 July 1995
419	Delayed Neurotoxicity of Organophosphorus Substances: 29-Day Repeated Dose Study	4 April 1984	1		27 July 1995
420	Acute Oral toxicity—Fixe Dose Procedure	17 July 1992	1		17 December 2001
421	Reproduction/ Developmental Toxicity Screening Test	27 July 1995	2	28 July 2015	26 July 2016
422	Combined Repeated Dose Toxicity Study with the Reproduction/ Developmental Toxicity Screening Test	22 March 1996	2	28 July 2015	26 July 2016
423	Acute Oral Toxicity—Acute Toxic Class Method	22 March 1996	1		17 December 2001
424	Neurotoxicity Study in Rodents	21 July 1997	0		–
425	Acute Oral Toxicity: Up-and- Down Procedure	21 September 1998 (corrected in2008)	2	17 December 2001	23 March 2006
426	Developmental Neurotoxicity Study	16 October 2007	0		–
427	Skin Absorption: in vivo Method	13 April 2004			–
428	Skin Absorption: in vitro Method	13 April 2004			–
429	Skin Sensitisation: Local Lymph Node Assay	24 April 2002	1		22 July 2010

(continued)

Table 1.8 (continued)

No.	Title	Original adoption	No. of updates	Updates	Most recently updated
430	in vitro Skin Corrosion: Transcutaneous Electrical Resistance Test (TER)	13 April 2004	2	26 July 2013	28 July 2015
431	in vitro Skin Corrosion: Reconstructed Human Epidermis (RHE) Test	13 April 2004	4	26 July 2013 26 Sept. 2014 28 July 2015	26 July 2016
432	in vitro 3T3 NRU Phototoxicity Test	13 April 2004	0		–
433	Acute Inhalation Toxicity test, Fixed Concentration Procedure	9 October 2017 (Corrected 27 June 2018)	0		–
435	in vitro Membrane Barrier Test Method for Skin Corrosion	19 July 2006	1		28 July 2015
436	Acute Inhalation Toxicity—Acute Toxic Class Method	7 September 2009	0		–
437	Bovine Corneal Opacity and Permeability Test Method for Identifying Ocular Corrosives and Severe Irritants	7 September 2009 (corrected in2010) (corrected in2017)	1		26 July 2013
438	Isolated Chicken Eye Test Method for Identifying Ocular Corrosives and Severe Irritants	7 September 2009 (corrected in 2017)	2	26 July 2013	27 June 2018
439	in vitro Skin Irritation: Reconstructed Human Epidermis Test Method	22 July 2010	2	26 July 2013	28 July 2015
440	Uterotrophic Bioassay in Rodents: A short-term screening test for oestrogenic properties	16 October 2007	0		–
441	Hershberger Bioassay in Rats: A short-term Screening Assay for	7 September 2009	0		–

(continued)

Table 1.8 (continued)

No.	Title	Original adoption	No. of updates	Updates	Most recently updated
	(Anti)Androgenic Properties				
442A	Skin Sensitisation: Local Lymph Node Assay: DA	22 July 2010	0		–
442B	Skin Sensitisation: Local Lymph Node Assay: BrdU-ELISA	22 July 2010	1		27 June 2018
442C	In chemico skin Sensitisation	5 February 2015	0		–
442D	In vitro Skin Sensitisation	5 February 2015	1		27 June 2018
442E	In vitro Skin Sensitisation: In Vitro Skin Sensitisation assays addressing the Key Event on activation of dendritic cells on the Adverse Outcome Pathway for Skin Sensitisation	26 July 2016	1		9 October 2017 (Corrected 27 June 2018)
443	Extended One-Generation Reproductive Toxicity Study	28 July 2011 (corrected in2012) (corrected in2018)	0		
451	Carcinogenicity Studies	12 May 1981	1		7 September 2009 (corrected in2018)
452	Chronic Toxicity Studies	12 May 1981	1		7 September 2009 (corrected in 2018)
453	Combined Chronic Toxicity/ Carcinogenicity Studies	12 May 1981	1		7 September 2009 (corrected in 2018)
455	Performance-based Test Guideline for Stably Transfected Transactivation in vitro Assays to Detect Estrogen Receptor Agonists and Antagonists	7 September 2009	3	2 October 2012 28 July 2015	26 July 2016

(continued)

Table 1.8 (continued)

No.	Title	Original adoption	No. of updates	Updates	Most recently updated
456	H295R Steroidogenesis Assay	28 July 2011	0		–
457	BG1Luc Estrogen Receptor Transactivation in vitro Assay to Detect Estrogen Receptor Agonists and Antagonists	2 October 2012	0		Date of deletion 29 January 2018
458	Stably Transfected Human Androgen Receptor Transcriptional Activation Assay for Detection of Androgenic Agonist and Antagonist Activity of Chemicals	26 July 2016	0		–
460	Fluorescein Leakage Test Method for Identifying Ocular Corrosives and Severe Irritants	2 October 2012 (corrected in 2017)	0		
471	Bacterial Reverse Mutation Test	26 May 1983	1		21 July 1997
472	Genetic Toxicology: *Escherichia coli*, Reverse Assay	26 May 1983	0		Date of deletion: 21 July 1997 (Method merged with TG 471)
473	In vitro Mammalian Chromosome Aberration Test	26 May 1983 (corrected in 2016)	2	21 July 1997	26 September 2014
474	In vivo Mammalian Erythrocyte Micronucleus Test	26 May 1983 (corrected in2016)	2	21 July 1997	26 September 2014
475	In vivo Mammalian Bone Marrow Chromosome Aberration Test	4 April 1984 (corrected in 2016)	2	21 July 1997	26 September 2014
476	In vitro Mammalian Cell Gene Mutation Tests using the Hprt and xprt genes	4 April 1984 (corrected in 2016)	2	21 July 1997	28 July 2015

(continued)

Table 1.8 (continued)

No.	Title	Original adoption	No. of updates	Updates	Most recently updated
477	Genetic Toxicology: Sex- Linked Recessive Lethal Test in *Drosophilia melanogaster*	4 April 1984	0		Date of deletion 2 April 2014
478	Genetic Toxicology: Rodent Dominant Lethal Test	4 April 1984 (corrected in 2016)	1		28 July 2015
479	Genetic Toxicology: in vitro Sister Chromatid Exchange assay in Mammalian Cells	23 October 1986	0		Date of deletion 2 April 2014
480	Genetic Toxicology: *Saccharomyces cerevisiae*, Gene Mutation Assay	23 October 1986	0		Date of deletion 2 April 2014
481	Genetic Toxicology: *Saccharomyces cerevisiae*, Mitotic Recombination Assay	23 October 1986	0		Date of deletion 2 April 2014
482	Genetic Toxicology: DNA Damage and Repair, Unscheduled DNA Synthesis in Mammalian Cells in vitro	23 October 1986	0		Date of deletion 2 April 2014
483	Mammalian Spermatagonial Chromosomal Aberration Test	23 October 1986 (corrected in 2016)	2	21 July 1997	28 July 2015
484	Genetic Toxicology: Mouse Spot Test	23 October 1986	0		Date of deletion 2 April 2014
485	Genetic Toxicology: Mouse Heritable Translocation Assay	23 October 1986	0		–
486	Unscheduled DNA Synthesis (UDS) Test with Mammalian Liver Cells in vivo	21 July 1997	0		–
487	In vitro Mammalian Cell Micronucleus Test	22 July 2010 (corrected in 2016)	1		26 September 2014
488	Transgenic Rodent Somatic and Germ	28 July 2011 (corrected in 2016)	1		26 July 2013

(continued)

Table 1.8 (continued)

No.	Title	Original adoption	No. of updates	Updates	Most recently updated
	Cell Gene Mutation Assays				
489	In vivo Mammalian Alkaline Comet Assay	22 September 2014	0		–
490	In vitro Thymidine Kinase Mutation Test	28 July 2015 (corrected in 2016)	0		
491	Short-time Exposure for the Detection of Chemicals Causing Serious Eye Damage, and Chemicals Not Requiring Classification for Serious Eye Damage or Eye Irritation	28 July 2015 (corrected in2017) (Corrected 27 June 2018)	0		–
492	Reconstructed Human Cornea-like Epithelium for the Detection of Chemicals Not Requiring Classification and Labelling for Eye Irritation or Serious Eye Damage	28 July 2015 (corrected in 2017)	1		27 June 2018
493	Human Recombinant Estrogen Receptor Binding Assay	28 July 2015	0		–

1.6.2.2 ICH Coordinators

The ICH Coordinator acts as the main contact point for the ICH Secretariat. Coordinators are nominated by the six co-sponsoring parties and they help in the smooth conduction of ICH functions.

1.6.2.3 ICH Secretariat

The Secretariat is responsible for meeting preparations and documentation and coordinating the Working Group and Discussion Group meetings.

1.6.2.4 ICH Working Group

The Steering committee supports the formation of working groups, namely, an Expert Working Group (EWG), an Implementation Working Group (IWG) or an Informal Working Group, depending on the type of harmonisation activity.

Table 1.9 Section 5—Other Test guidelines [Adapted from (OECD Guidelines for the Testing of Chemicals, Section 5 n.d., p. 5)]

No.	Title	Original adoption	No. of updates	Updates	Most recently updated
501	Metabolism in Crops	10 January 2007	0		–
502	Metabolism in Rotational Crops	10 January 2007	0		–
503	Metabolism in Livestock	10 January 2007	0		–
504	Residues in Rotational Crops (Limited Field Studies)	10 January 2007	0		–
505	Residues in Livestock	10 January 2007	0		–
506	Stability of Pesticide Residues in Stored Commodities	16 October 2007	0		–
507	Nature of the Pesticide Residues in Processed Commodities—High-Temperature Hydrolysis	16 October 2007	0		–
508	Magnitude of Pesticide Residues in Processed Commodities	3 October 2008	0		–
509	Crop Field Trial	7 September 2009	0		–

1.6.3 ICH Guidelines (ICH Official web site: ICH n.d.)

ICH developed guidelines on the four major domains (Tables 1.10, 1.11, 1.12, and 1.13) viz. Quality, Efficacy, Safety and Multidisciplinary.

1.7 Environmental Protection Agency (EPA)

The Environmental Protection Agency (EPA) is a US federal government agency established by the executive order of the then US President Richard Nixon in Dec 1970. The EPA headquarters is located in Wexford and regional inspectorates in Dublin, Cork, Kilkenny, Castlebar and Monaghan and the regional offices in Athlone and Limerick.

It was primarily developed to safeguard human and environmental health. The EPA was also assigned to fix safety limits for chemicals and pollutants in food, animal feed and water.

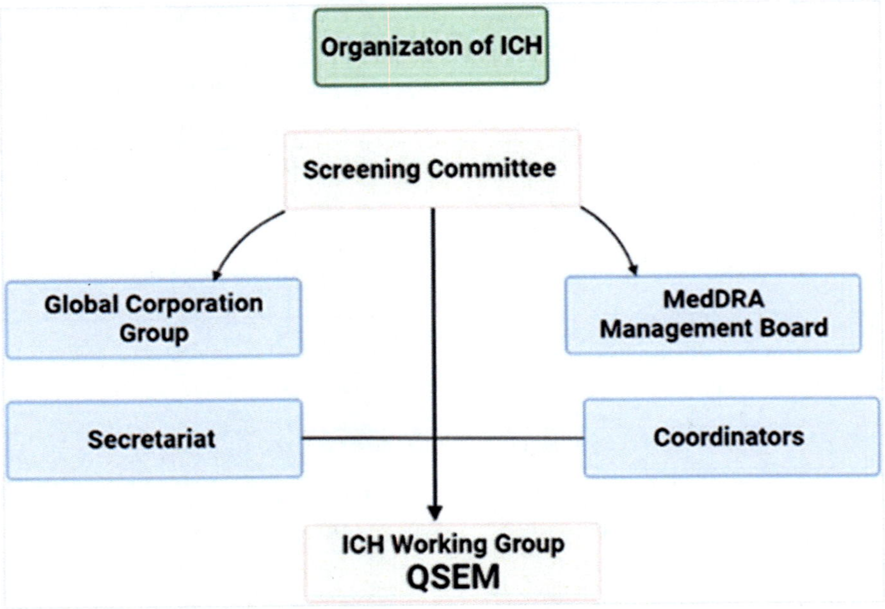

Fig. 1.2 ICH organisational structures. [Adapted from (ICH Official web site: ICH n.d.)]

1.7.1 EPA Organization Structure (Fig. 1.3)

A dedicated executive board comprising a Director-General and five Directors manage the functions of EPA. The activities were governed by the Directors in the five offices viz.

- The Office of Environmental Enforcement
- The Office of Environmental Sustainability
- The Office of Evidence and Assessment
- The Office of Radiation Protection and Environmental Monitoring
- The Office of Communications and Corporate Services

1.7.2 Functions

In addition to the primary responsibility to safeguard environmental and human health, the other functions of EPA include:

- Environmental licensing
- Enforcement of environmental law

Table 1.10 ICH Quality guidelines [Adapted from (ICH Official web site: ICH n.d.)]

Quality guidelines	
Q1A	Stability testing of new drug substances and products
Q1B	Stability testing: photo stability testing of new drug substances and products
Q1C	Stability testing for new dosage forms
Q1D	Bracketing and matrixing designs for stability testing of new drug substances and products
Q1E	Evaluation for stability data
Q1F	Stability data package or registration applications in climatic zones (iii) and (iv)
Q2(R1)	Validation of analytical procedures: text and methodology
Q2(R1)/Q14 EWG	Analytical Procedure Development and Revision of Q2(R1) Analytical?
Q3A(R2)	Impurities in new drug substances
Q3B(R2)	Impurities in new drug products
Q3C(R6)	Maintenance of guidelines for residual solvents
Q3C(R8)	EWG Maintenance of the guidelines for residual solvents
Q3D(R1)	Guideline for elemental impurities
Q3D(R2)	Revision of Q3D(R1) for cutaneous and transdermal products
Q3D	Implementation of guidelines for external impurities
Q3E	Impurity: assessment and control of extractable and leachable for pharmaceutical and biologics
Q4A	Pharmacopoeias harmonisation
Q4B	Evaluation and recommendation of pharmacopoeias texts for use in the ICH regions
Q4B Annex1 (R1)	Residue on ignition/sulphated ash general chapter
Q4B Annex2 (R1)	Test for extractable volume of parenteral preparations—general chapter
Q4B Annex3 (R1)	Test for particulate contamination: sub-visible particles—general chapter
Q4B Annex4A (R1)	Microbiological examination of non-sterile products: microbial enumeration test general chapter
Q4B Annex4B (R1)	Microbiological examination of non-sterile products: Test for specified micro-organisms general chapter
Q4B Annex4C (R1)	Microbiological examination of non-sterile products: acceptance criteria for pharmaceutical preparation and substances for pharmaceutical use general chapter
Q4B Annex5(R1	Disintegration test general chapter
Q4B Annex6	Uniformity of dosage units general chapter
Q4BAnnex7 (R2)	Dissolution test general chapter
Q4B Annex8 (R1)	Sterility test general chapter
Q4B ANNEX 9 (R1)	Tablet friability general chapter

(continued)

Table 1.10 (continued)

Quality guidelines	
Q4B ANNEX 10 (R1)	Polyacrylamide gel electrophoresis general chapter
Q4B ANNEX 11	Capillary electrophoresis general chapter
Q4B ANNEX 12	Analytical sieving general chapter
Q4B ANNEX 13	Bulk density and tapped density of powders general
Q4B ANNEX 14	Bacterial endotoxins test general chapter
Q4B FAQs	Frequently asked questions
Q5A(R1)	Viral safety evaluation of biotechnology products derived from cell lines of human or animal origin
Q5A(R2) EWG	Viral Safety Evaluation of Biotechnology Products Derived from Cell Lines of Human or Animal Origin
Q5B	Analysis of the expression construct in cells used for the production of r-DNA-derived protein products
Q5C	Stability testing of biotechnological/biological products
Q5D	Derivation and characterization of cell substrates used for the production of biotechnological/biological products
Q5E	Comparability of biotechnological/biological products subject to changes in their manufacturing process
Q6A	Specifications: Test procedures and acceptance criteria for new drug substances and new drug products: chemical substances
Q6B	Specifications: Test procedures and acceptance criteria for biotechnological/ biological products
Q7	Good manufacturing practice guide for active pharmaceutical ingredients
Q7 Q&As	Good Manufacturing Practice Guide for active pharmaceutical ingredients questions and answers
Q8(R2)	Pharmaceutical development
Q8/9/10 Q&As (R4)	Q8/Q9/Q10 Implementation
Q9	Quality risk management
Q9(R1)	Quality risk management
Q8/9/10 Q&As (R4)	Q8/9/10 Implementation
Q10	Pharmaceutical quality system
Q8/9/10 Q&As (R4)	Q8/Q9/Q10 Implementation
Q11	Development and manufacture of drug substances (chemical entities and biotechnological/biological entities)
Q11 Q&As	Questions & answers: Selection and justification of starting material for the manufacture of drug substances
Q12	Technical and regulatory considerations for pharmaceutical product lifecycle management
Q12 IWG	Training on regulatory and technical considerations pharmaceutical product life management
Q13	Continuous manufacturing for drug substances and drug products
Q2(R2)/Q14 EWG	Analytical procedure development and revision of Q2(R1) analytical validation

Table 1.11 ICH Safety guidelines [Adapted from (ICH Official: Safety guidelines n.d.)]

Safety guidelines	
S1A	Need for carcinogenicity studies of pharmaceuticals
S1B	Testing for the carcinogenicity of pharmaceuticals
S1C(R2)	Dose selection for carcinogenicity studies of pharmaceuticals
S1(R1)	Rodent carcinogenicity studies for human pharmaceuticals
S2(R1)	Guidance on genotoxicity testing and data interpretation for pharmaceuticals intended for human use
S3A	Note for guidance on toxicokinetics: The assessment of systemic exposure in toxicity studies
S3A Q&As	Questions and answers: note for guidance on toxicokinetics: the assessment of systemic exposure in toxicity studies focus on microsampling
S3B	Guidance for repeated dose tissue distribution studies
S4	Duration of chronic toxicity testing in animals (rodent and non-rodent toxicity testing)
S5 (R3)	Revision S5 of detection of reproductive and developmental toxicity for human pharmaceuticals
S5 (R4) Maintenance EWG	Revision S5 of detection of reproductive and developmental toxicity for human pharmaceuticals
S6(R1)	Preclinical safety evaluation of biotechnology-derived pharmaceuticals
S7A	Safety pharmacology studies for human pharmaceuticals
S7B	The non-clinical evaluation of the potential for delayed ventricular repolarization (QT interval prolongation) by human pharmaceuticals
E14/S7B IWG	Questions and Answers: Clinical and non-clinical evaluation of QT/QTc interval prolongation and pro-arrhythmic potential
S8	Immunotoxicity studies for human pharmaceuticals
S9	Non-clinical evaluation for anticancer pharmaceuticals
S9 Q/As	Questions and Answers: non-clinical evaluation for anticancer pharmaceuticals
S10	Photo safety evaluation of pharmaceuticals
S11	Non-clinical safety testing in support of the development of paediatric pharmaceuticals
S12 EWG	Non-clinical bio-distribution considerations for gene therapy products

- Environmental planning, education and guidance
- Monitoring, analysing and reporting on the environment
- Regulating Ireland's greenhouse gas emissions
- Environmental research development
- Strategic environmental assessment
- Waste management
- Radiological protection

Table 1.12 ICH Efficacy guidelines [Adapted from (ICH Official: Efficacy guidelines n.d.)]

Efficacy guidelines	
E1	The extent of population exposure to assess clinical safety for drugs intended for long-term treatment of non-life-threatening conditions
E2A	Clinical safety data management: definitions and standards for expedited reporting
E2B(R3)	Clinical safety data management: Data element for transmission of individual case safety reports (ICSRs)
E2B(R3) Q/As	Clinical safety data management: Data element for transmission of individual case safety reports
E2B(R3) EWG/ IWG	Electronic transmission of individual case safety reports (ICSRs)
E2C(R2)	Periodic benefit-risk evaluation report (PBRER)
E2C(R2)Q/As	Questions and Answers: periodic benefit-risk evaluation report
E2D	Post-approval safety data management: Definitions and standards for expedited reporting
E2D(R1) EWG	Post-approval safety data management: definitions and standards for expedited reporting
E3	Structure and content of clinical study reports
E3 Q&As (R1)	Questions and Answers: structure and content of clinical study reports
E4	Dose-response information to support drug registration
E5(R1)	Ethnic factors in the acceptability of foreign clinical data
E5 Q&As (R1)	Questions &Answers: Ethics factors in the acceptability of Foreign Clinical Data
E6(R2)	Good clinical practice (GCP)
E6(R3) EWG	Good clinical practice (GCP)
E7	Studies in support of special populations: Geriatrics
E7 Q&As	Questions & Answers: Studies in support of special populations: Geriatrics
E8	General considerations for clinical trials
E8(R1) EWG	Revision on general considerations for clinical studies
E9	Statistical principles for clinical trials
E9(R1) EWG	Addendum: statistical principles for clinical trials
E10	Choice of the control group and related issues in clinical trials
E11 (R1)	Addendum: clinical investigation of medicinal products in the paediatric population
E11A EWG	Paediatric extrapolation
E12A	Principles for clinical evaluation of new anti-hypertensive drugs
E14	The clinical evaluation of QT/QTc interval prolongation and proarrhythmic potential for non-antiarrhythmic drugs
E14 Q&As (R3)	Questions & Answers: The clinical evaluation of QT/QTc interval prolongation and pro-arrhythmic potential for non-antiarrhythmic drugs
E14/S7B IWG	Questions and Answers: clinical and non-clinical evaluation of QT/QTc interval prolongation and proarrhythmic potential
E15	Definitions for genomic biomarkers, pharmacogenomics, pharmacogenetics, genomic data and sample coding categories
E16	Biomarkers related to drug or biotechnology product development: context, structure and format of qualification submissions

(continued)

Table 1.12 (continued)

Efficacy guidelines	
E17	General principles for planning and design of multi-regional clinical trials
E18	Genomic sampling and management of genomic data
E19	Optimization of safety data collection
E20 EWG	Adaptive clinical trials

Table 1.13 ICH Multidisciplinary guidelines [Adapted from (ICH Official: Multidisciplinary guidelines n.d.)]

Multidisciplinary guidelines	
M1	MedDRA-Medical Dictionary for Regulatory Activities
M1 PtC WG	MedDRA Points to Consider
M2 EWG	Electronic standards for the transfer of regulatory information
M3(R2)	Non-clinical safety studies for the conduct of human clinical trials and marketing authorization for pharmaceuticals
M3(R2) Q&As(R2)	Questions and Answers: Guidance on non-clinical safety studies for the conduct of human clinical trials and marketing authorisation for pharmaceuticals
M4	The common Technical Document (cTc)
M5	Data elements and standards for drug dictionaries
M6	Virus and gene therapy vector shedding and transmission
M7(R1)	Assessment and control of DNA reactive (mutagenic) impurities in pharmaceuticals to limit potential carcinogenic risk
M7(R2) EWG/ IWG	Assessment and control of DNA reactive (mutagenic) impurities in pharmaceuticals to limit potential carcinogenic risk
M8(eCTD) v3.2.2	Electronic Common Technical Document—(eCTD) v3.2.2
M8(ectd) v4.0	Electronic Common Technical Document—(eCTD) v4.0
M8 EWG/IWG	Electronic Common Technical Document—(eCTD)
M9	Biopharmaceutics classification system-based biowaivers
M9 Q&AS	Q&AS on biopharmaceutics classification system-based biowaivers
M10 EWG	Bioanalytical method validation
M11 EWG	Clinical electronic Structured Harmonised Protocol (CeSHarP)
M12 EWG	Drug Interaction Studies
M13	Bioequivalence for Immediate-Release Solid Oral Dosage Forms

1.7.3 EPA Health Effects Test Guidelines

Under FIFRA and TSCA, EFA health effects testing guidelines are designed to meet the testing criteria for human health impacts of chemical compounds (Table 1.14, 1.15, 1.16, 1.17, 1.18, 1.19, and 1.20).

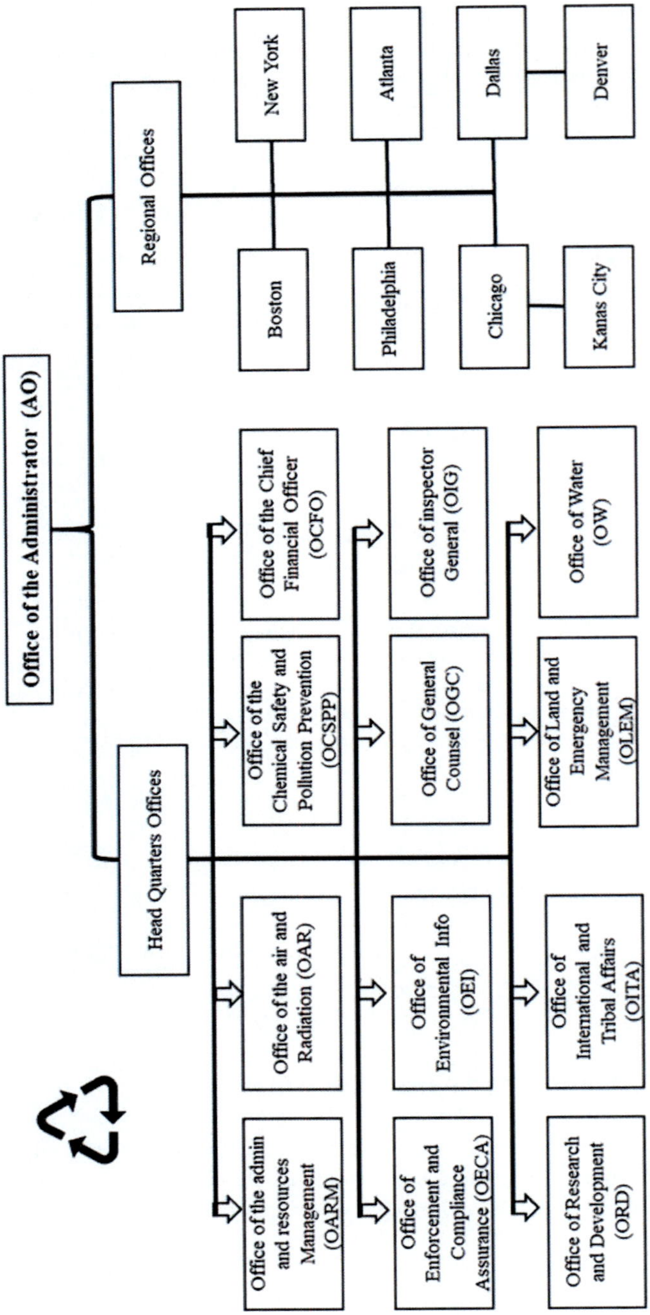

Fig. 1.3 EPA organisational structure. [Adapted from (epa-org-chart n.d.)]

Table 1.14 Section-A (Acute Toxicity Test guidelines) [Adapted from (US EPA 2015)]

Guideline no	Title	Updates
870.1000	Acute Toxicity Testing Background	December 2002
870.1100	Acute Oral Toxicity	December 2002
870.1200	Acute Dermal Toxicity	August 1998
870.1300	Acute Inhalation Toxicity	August 1998
870.2400	Acute Eye Irritation	August 1998
870.2500	Acute Dermal Irritation	August 1998
870.2600	Skin Sensitization	March 2003

Table 1.15 Group-B—Subchronic Toxicity Test guidelines [Adapted from (US EPA 2015)]

Guideline no	Title	Updates
870.3050	Repeated Dose 28-Day Oral Toxicity Study in Rodents	July 2000
870.3100	90 Day Oral Toxicity in Rodents	August 1998
870.3150	90 Day Oral Toxicity in Non-rodents	August 1998
870.3200	21/28 Day Dermal Toxicity	August 1998
870.3250	90 Day Dermal Toxicity	August 1998
870.3465	90 Day Inhalation Toxicity	August 1998
870.3550	Reproduction/Developmental Toxicity Screening Test	July 2000
870.3650	Combined Repeated Dose Toxicity Study with the Reproduction/ Developmental Toxicity Screening Test	July 2000
870.3700	Prenatal Developmental Toxicity Study	August 1998
870.3800	Reproduction and Fertility Effects	August 1998

Table 1.16 Group-C—Chronic Toxicity Test guidelines [Adapted from (US EPA 2015)]

Guideline no	Title	Updates
870.4100	Chronic Toxicity	August 1998
870.4200	Carcinogenicity	August 1998
870.4300	Combined Chronic Toxicity/Carcinogenicity	August 1998

Table 1.17 Group-D—Genetic Toxicity Test guidelines [Adapted from (US EPA 2015)]

Guideline no	Title	Updates
870.5100	Bacterial Reverse Mutation Test	August 1998
870.5140	Gene Mutation in Aspergillus nidulans	August 1998
870.5195-	Mouse Biochemical Specific Locus Test	August 1998
870.5200	Mouse Visible Specific Locus Test	August 1998
870.5250	Gene Mutation in Neurosporacrassa	August 1998
870.5275	Sex-linked Recessive Lethal Test in Drosophila melanogaster	August 1998
870.5300	In vitro Mammalian Cell Gene Mutation Test	August 1998
870.5375	In vitro Mammalian Chromosome Aberration Test	August 1998
870.5380	Mammalian Spermatogonial Chromosomal Aberration Test	August 1998
870.5385	Mammalian Bone Marrow Chromosomal Aberration Test	August 1998
870.5395	Mammalian Erythrocyte Micronucleus Test	August 1998
870.5450	Rodent Dominant Lethal Assay	August 1998
870.5460	Rodent Heritable Translocation Assays	August 1998
870.5500	Bacterial DNA Damage or Repair Tests	August 1998
870.5550	Unscheduled DNA Synthesis in Mammalian Cells in Culture	August 1998
870.5575	Mitotic Gene Conversion in Saccharomyces cerevisiae	August 1998
870.5900	In vitro Sister Chromatid Exchange Assay	August 1998
870.5915	In vivo Sister Chromatid Exchange Assay	August 1998

Table 1.18 Group-E—Neurotoxicity Test guidelines [Adapted from (US EPA 2015)]

Guideline no	Title	Updates
870.6100	Acute and 28-Day Delayed Neurotoxicity of Organophosphorus Substances	August 1998
870.6200	Neurotoxicity Screening Battery	August 1998
870.6300	Developmental Neurotoxicity Study	August 1998
870.6500	Schedule-Controlled Operant Behaviour	August 1998
870.6850	Peripheral Nerve Function	August 1998
870.6855	Neurophysiology Sensory Evoked Potentials	August 1998

Table 1.19 Group-F—Special Studies Test guidelines [Adapted from (US EPA 2015)]

Guideline no	Title	Updates
870.7200	Companion Animal Safety	August 1998
870.7485	Metabolism and Pharmacokinetics	August 1998
870.7600	Dermal Penetration	August 1998
870.7800	Immunotoxicity	August 1998

Table 1.20 Group-G—Health Effects Chemical-Specific Test guidelines [Adapted from (US EPA 2015)]

Guideline no	Title	Updates
870.8355	Combined Chronic Toxicity/Carcinogenicity Testing of Respirable Fibrous Particles	July 2001

1.8 Central Drugs Standard Control Organisation

The Central Drugs Standard Control Organisation (CDSCO) is the national regulatory body, which functions under the gamut of Directorate General of Health Services, Ministry of Health & Family Welfare, Government of India. CDSCO functions following the Drugs and Cosmetics Act, 1940 and Drugs and Cosmetics Rules 1945. The primary objectives of CDSCO is to ensure safer, effective and quality medicines, cosmetics, diagnostics and devices available to Indian citizens.

1.8.1 CDSCO Organisation System

The Drug Controller General of India is the head of CDSCO and the headquarters is situated at FDA Bhawan, New Delhi (Fig. 1.4]).

1.8.2 Location of CDSCO and Associated Offices

CDSCO has six zonal offices, four sub zonal offices, 13 Port offices and seven laboratories spread across the country (Table 1.21).

1.8.3 Functions of CDSCO

- Approval of new drugs and clinical trials
- Import Registration and licensing
- License approval of Blood Banks, LVPs, Vaccines, r-DNA products and some Medical Devices (CLAA Scheme)
- Amendment of D & C Act and Rules
- Banning of drugs and cosmetics
- Grant of Test License, Personal License, NOCs for Export
- Testing of New Drugs
- Oversight and market surveillance through the inspectorate of Centre, over and above the State Authority

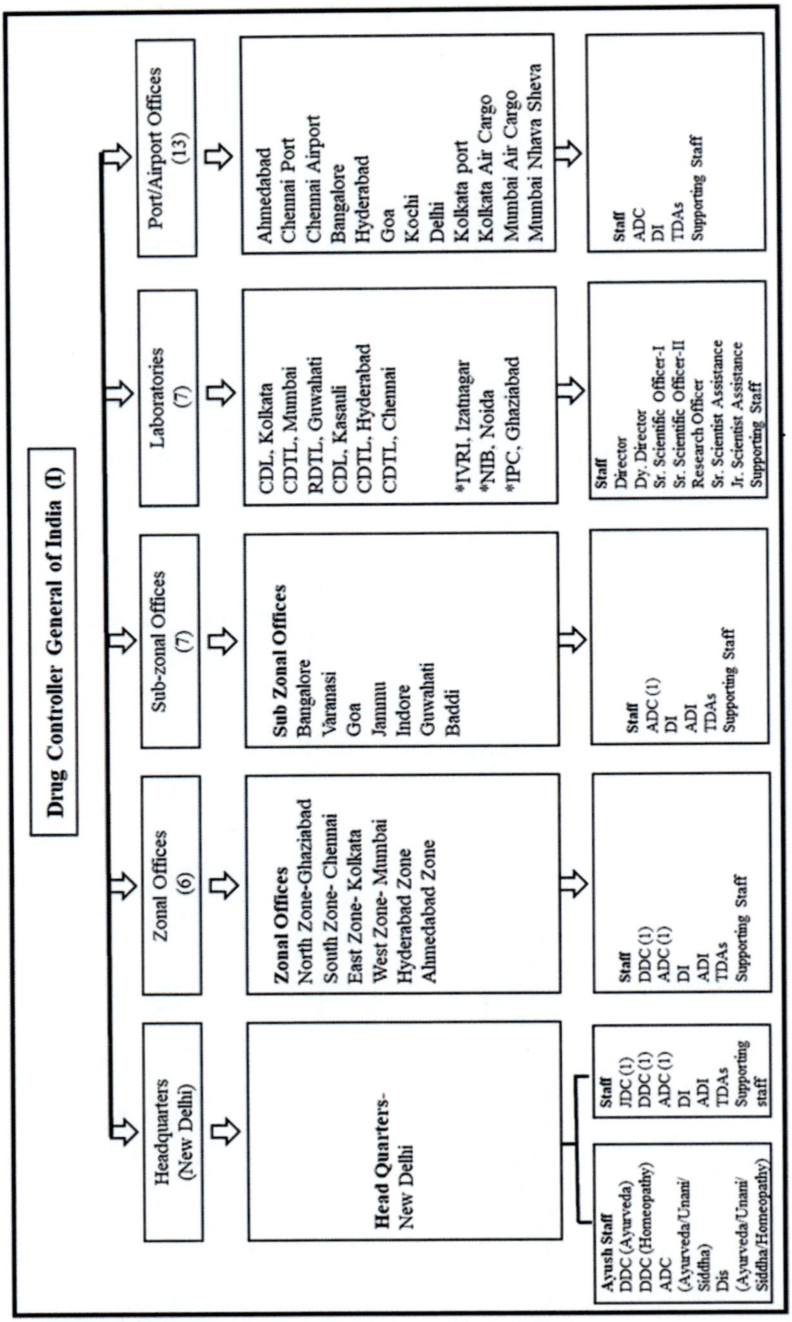

Fig. 1.4 CDSCO Organisation Chart. [Adapted from (CDSCO n.d.)]

Table 1.21 Location of CDSCO and associated offices (CDSCO n.d.)

Offices	Locations
Zonal Offices (6)	• North zone—Ghaziabad • South zone—Chennai • East zone—Kolkata • West zone—Mumbai • Hyderabad zone • Ahmedabad zone
Sub-zonal offices (7)	• Bangalore • Varanasi • Goa • Jammu • Indore • Guwahati • Baddi
Port offices/Airports (13)	• Ahmedabad • Chennai port • Chennai Airport • Bangalore • Hyderabad • Goa • Kochi • Delhi • Kolkata port • Kolkata Air Cargo • Mumbai Air Cargo • Mumbai NhavaSheva • Mumbai Custom House
Laboratories (7)	• CDL, Kolkata • CDTL,Mumbai • RDTL, Guwahati • CDL, Kasauli • CDTL, Hyderabad • CDTL, Chennai • *IVRI Izatnagar • *NIB Noida • *IPC Ghaziabad

1.9 Functions of State Licensing Authority

- Licensing of drug, API and finished formulations manufacturing site
- Licensing of establishments for sale or distribution of drugs
- Licensing of drug testing laboratories
- Monitoring the quality of drugs and cosmetics in the market

- Investigation and prosecution of contravention of legal provisions
- Removal of substandard drugs in the market

1.10 Schedule Y (Schedule Y(amended version) – CDSCO n.d.)

Schedule Y is a part of X-A of Drugs and Cosmetics Act 1945. It provides information on the data required for approval of clinical trials and import or manufacture of a new drugs for marketing in India.

1.10.1 Scope

- Allopathic drug
- Ayurvedic, Siddha, Homeopathic and Unani Drugs
- Blood and Blood products
- Medical devices (Tables 1.22 and 1.23)

1.10.2 Non-Clinical Toxicity studies as per the Schedule-Y (Appendix—III)

General Aspects

The studies should be conducted following Schedule-Y standard protocols or guidelines. The test item and test system should be characterised before the initiation of the study. All documents and materials including study plan, raw data, formats, draft and final study reports, paraffin blocks and histological slides should be archived for minimum of 5 years following the introduction of the drug in the market.

1.10.2.1 Systemic Toxicity Studies (Schedule Y(amended version) - CDSCO n.d.)

a. Single-Dose Toxicity Studies

As per Schedule-Y, an acute (single-dose) study should be carried out at least in two species (5 animals of either sex) viz. one rodent and a non-rodent model using the

Table 1.22 Schedule Y rules [Adapted from (Imran et al. 2013) with modifications]

Rule	Details
122A	Application for permission to import new drug
122B	Application for approval to manufacture a new drug
122D	Permission to import or manufacture FDC.
122DA	Permission to conduct clinical trials for new drug/investigational new drug.
122DAA	Clinical Trial Definitions
122E	New Drug Definitions

Table 1.23 Schedule appendices (Schedule Y(amended version) – CDSCO n.d.)

Appendix—I	Data to be submitted along with the application to conduct clinical trials/import/ manufacture of new drugs for marketing in the country
Appendix—IA	Data required to be submitted by an applicant for grant of permission to import and/or manufacture a new drug already approved in the country
Appendix—II	Structure, Contents and Format for Clinical Study Reports
Appendix—III	Animal Toxicology (Non-Clinical Toxicity studies)
Appendix—IV	Animal Pharmacology
Appendix—V	Informed Consent
Appendix—VI	Fixed dose Combinations (FDC's)
Appendix—VII	Undertaking by Investigator
Appendix—VIII	Ethics Committee
Appendix—IX	Stability testing of New Drugs
Appendix—X	Contents of the proposed protocol for conducting clinical trials
Appendix—XI	Data elements for reporting serious adverse events occurring in a Clinical Trial

Table 1.24 Species for the repeated day toxicity study

	Species	
Study	Rodent	Non-rodent
14/28-day repeated-dose toxicity studies	6-10/sex/group	2-3/sex/group
90-day repeated-dose toxicity studies	15-30/sex/group	4-6/sex/group
180-day repeated-dose toxicity studies	15-30/sex/group	4-6/sex/group

same route intended for human exposure. Unless the route is intravenous, one more route of administration should be performed to ensure the systemic absorption of the test item. A limit of 2000 mg/kg is suggested for oral dosing. Following test item dosing, the animals should be monitored for death, onset and disappearance of toxic clinical signs. Further, gross pathology should be performed and microscopic findings as required. The minimum lethal dose (MLD) and maximum tolerated dose (MTD) are established by observing 14 days after the dosing.

b. Repeated-Dose Systemic Toxicity Studies

One rodent and one non-rodent species should be used in a repeated dose systemic toxicity study (Table 1.24). The study should be performed for 14, 28, 90 or 180-days, as per the therapeutic indications and scale of the planned clinical trial (s). The route of drug administration should be identified as intended for human exposure. The highest dose should be selected in such a way that the test item does not produce mortality and the lowest dose should produce observable levels of toxicity signs.

Observations include physiological, behavioural, biochemical, gross pathology, haematological and microscopical examination of tissue sections. In case of parenteral drugs, the injection site should be examined for gross and microscopical changes. In non-rodent species, the initial and final electrocardiogram, as well as fundus examination, is recommended. (ICH Topic M 3 (R2) Non-Clinical Safety Studies for the Conduct of Human Clinical Trials and Marketing Authorization for Pharmaceuticals - Google Search n.d.)

End-point parameters

Haematological: Haematocrit (HCT), haemoglobin (HGB), total erythrocyte count (RBC), mean corpuscular haemoglobin (MCH), mean corpuscular volume (MCV), mean corpuscular haemoglobin concentration (MCHC), platelet count, total leucocytes (WBC), blood clotting time, reticulocytes, neutrophils, lymphocytes, eosinophils basophils and monocytes, ESR (Non-rodents only)

Coagulation Parameters: (Non-rodents only): Bleeding, coagulation, prothrombin, activated partial thromboplastin-time

Urine parameters: Colour, specific gravity, volume, pH, albumin, glucose, ketone bodies, bile pigments, urobilinogen, blood cells

Biochemical parameters: Glucose, cholesterol, triglycerides, total protein, albumin, serum glutamyl pyruvate aminotransferase (SGPT), γ-glutamyl transferase (SGOT), total bilirubin, total bile acid creatinine and urea, Blood Urea Nitrogen, alkaline phosphate (ALP), GGT (Non-rodents only), globulin

Serum electrolytes: Total calcium, potassium, sodium, chloride, inorganic calcium

c. Male Fertility Study (Schedule Y(amended version) - CDSCO n.d.)

The study is performed in only one species, preferably rats (6 males/group), unless the criteria mandate otherwise. The dose for the male fertility study should be fixed based on the results from the 14/28-day repeated dose study. The experimental animals should be pre-treated with the test item for a period of 28–70 days before introducing mating (1 male: 2 females) and this to study the effects of the test item on male fertility. The treatment should be continued during the mating period till the detection of vaginal plug or 10 days after introduction to mating, whichever is earlier. Pregnant females should be determined for their fertility index after day 13 of gestation. All the males should be sacrificed, and the weights of testes and epididymis should be taken. Sperms should be evaluated for motility and morphology.

d. Female Reproduction and Developmental Toxicity Studies

Segment I, II and III studies should be performed for the test item meant for administration during the childbearing age and pregnancy. These studies should be performed in rodents and segment II study should be performed using rabbits as non-rodent species.

1. Female Fertility Study (Segment I)

In the segment I study, one species is used, preferably rats. The test item should be given to both (15/group) males and (15/group) females (28 days in males and 14 days in females), before introducing for mating. Based on the MTD results obtained from previous systemic toxicity studies, three graded doses should be used usually. The route of administration should be the same as intended for clinical exposure. Body weight, feed intake, clinical signs of toxicity, mating behaviour, length of gestation, parturition and gross pathology and histopathology of affected organs of dams should be recorded. Pups from both the vehicle and test item treated groups should be observed for body weight, growth, sex-wise distribution, survival, any signs of toxicity, gross pathological examination and histopathology of organs showing lesions in gross pathology examination.

2. Teratogenicity Study (Segment II)

One rodent (preferably rat) and one non-rodent (rabbit) species should be used in the segment II investigation. The test item should be administered at three dose levels during the organogenesis phase. Minimum 20 pregnant rats or mice and 12 rabbits in control and treated groups. Observations include examination of the ovaries, uterus and uterine contents, number of corpora lutea, implantation sites and resorptions. In foetuses: total numbers, body weight and length, sex, feed intake, gross and histopathological examination.

3. Perinatal Study (Segment III)

Rat is the preferred species for segment III studies. The grouping includes one vehicle control and three test groups (three different dose levels) with 15 dams in each. The test item should be given until the end of the third trimester of pregnancy (day 15 of pregnancy), and the dose that causes minimal foetal loss should be continued during lactation and weaning. Dams should then be sacrificed and the parameters listed below.

One male and one female from the F1 generation (total of 15 males and 15 females in each group) should be selected and treated with a vehicle or test item. F1 generation's mating performance and fertility should be assessed in order to produce the F2 generation and the growth parameters should be monitored until weaning. Animals from F1/F2 generation treated with vehicle/test item should be sacrificed at the end of the study and analysed for parameters such, Dams: body weight, feed intake, signs of toxicity, the progress of gestation/parturition periods and gross pathology. In pups: clinical signs, sex-wise distribution, body weight, growth parameters, gross and histopathology examination should be performed.

1.10.2.2 Local Toxicity

Test item administered other than oral route needs to be evaluated for the local toxicity potential, preferably in two species.

a. Dermal Toxicity Study

Rabbits and rats are suggested for dermal toxicity studies. The test item should be applied on the shaven skin, covering not less than 10% of the total body surface area. The area must be covered intact using dressing gauge materials. The duration of administration depends on the clinical use period (7–90 days). The test item should be applied at four levels of concentration and inclusion of satellite groups is also recommended. Clinical signs such as erythema, oedema and eschar formation, as well as histological examination at the sites of application, should be used for scoring the irritation or lesions if any.

b. Photoallergic or Dermal Phototoxicity

The test is meant for the test item or its metabolite that causes photoallergic or dermal phototoxicity. The test is performed following the Armstrong and Harber method using guinea pigs. In the pre-test, the test item is applied on the animals ($n = 8$) at four levels of concentration (patch application for 2 h ± 15 min) with and without UV exposure (10 J/cm^2). The observations are recorded at 24 h and 48 h and the highest non-irritant dose should be determined. The main test should be performed with $n = 10$ in the test group and 5 in the control group. Animals should be challenged with the same concentration of test item with a similar 2 h application, followed by exposure to 10 J/cm^2 UV radiation between days 20 and 24 of the test. A reference control group might be added using compounds like musk ambrett or psoralin. The dermal reactions should be graded and scored.

c. Vaginal Toxicity Study

Rabbits or dogs (6–10/dose) are preferred species. The test item should be applied topically (vaginal mucosa) for a minimum of 7 days in the form of a pessary, cream or ointment. Observations include swelling, closure of introitus and histopathology examination of the vaginal wall.

d. Rectal Tolerance Test

Rabbits or dogs (6–10/dose) are recommended and the test item is applied at least for7 days. Observation includes clinical signs of toxicity, pain symptoms, blood and/or mucus in faeces, anal region/sphincter condition, gross and histopathological evaluation of rectal mucosa.

1.10.2.3 Parenteral Drugs

For the test item intended for intramuscular or intravenous or subcutaneous or intradermal injections, the sites of injection should be specifically evaluated both by gross and histopathological examination. The reversibility of adverse effects should be evaluated on a case-by-case basis, if necessary.

1.10.2.4 Ocular Toxicity Studies

Two species should be used to evaluate the ocular toxicity of the test item, one of which should be the rabbit. Liquids, gels or drug-saturated contact lenses should be used for a maximum of 90 days, unless the duration is specified. One eye serves as the test and the other one as the control. Ophthalmological examination (using slit-lamp) should be performed to detect the changes in the cornea, iris and aqueous humour. Fluorescent dyes (sodium fluorescein, 0.25–1.0% w/v) should be used for identifying the defects in the surface epithelium of the cornea and conjunctiva. Changes in intraocular pressure should be monitored using a tonometer. The histological examination of the eyes must be done at the end of the experiment.

1.10.2.5 Inhalational Toxicity Studies

The choice of species for inhalational studies includes one rodent and one non-rodent. Gases and vapours (particle size of 4 μm, in case of aerosol) should be given in whole-body exposure chambers (limit dose of 5 mg/L) for a maximum of 6 h per day and 5 days a week. The effects of the test item on respiratory rate and bronchial lavage fluid composition. Histopathological evaluation of respiratory passages and lung tissue should be considered along with the regular parameters of systemic toxicity studies.

1.10.2.6 Allergenicity/Hypersensitivity

Allergenicity/hypersensitivity tests include guinea pig maximisation test (GPMT) and local lymph node assay (LLNA).

a. Guinea Pig Maximization Test

The test should be performed in two steps viz.

1. Pre-test: To determine the maximum and the minimum irritant dose.
2. Main study: At least six male and six female animals per group should be used in the main study. Intradermal induction (day 1) along with epidermal challenge (day 21) is performed. To examine the intradermal induction dose (mild to moderate irritations), four dose levels should be tested by the intradermal injection of the test item in four male and four female animals (two animals of each sex should also be given the Freund's adjuvant). Similarly, the minimum irritant dose on epidermal injection (topical) should be determined in two males and two females, as challenge dosing. A re-challenge should be done 7–30 days following the first challenge if there is no response. The observations include erythema and oedema (individual animal scores as well as maximum grading) scoring.

b. Local Lymph Node Assay

The tests determined the skin sensitisation potential using the mouse model. Mice (6/group; either male or female) is used for local lymph node assay. The test item is given in the ear for three consecutive days. On day 5, flowing auricular lymph nodes should be dissected out 5 h after i.v. H-thymidine or bromo-deoxy-uridine (BrdU) injection. An increase in H-thymidine or Brd Userves as an indicator for skin sensitisation.

1.10.2.7 Genotoxicity

Genotoxic compounds are substances that impart mutation or toxicity to genetic materials causing hazards to humans. The genotoxicity tests enable hazard identification, concerning damage to DNA and its fixation.

The standard battery of tests is as follows

1. A test for bacterial gene mutation.
2. An in vitro test using mammalian cells for cytogenetic evaluation of chromosomal damage or an in vitro mouse lymphoma TK assay (MLA).
3. Chromosomal damage using rodent hematopoietic cells.

In-vitro studies include AMES test and chromosomal aberrations (CA) using cultured cell lines. In vivo studies include micronuclei using mice bone marrow cells. Before phase III clinical trials, the genotoxicity data collection should be completed.

1. In-vitro exposure (with and without metabolic activation) should be done at a minimum of 5 log dose levels. The test is usually performed using bacterial strain: *S. typhimurium* TA1535, TA1537 or TA97 or TA97a, TA98 and TA100, TA102 and *E. coli* WP2 uvrA, or E. coli WP2 uvrA (pKM101). The test should include positive control groups (reference compounds: 9-aminoacridine, 2-nitrofluorine, sodium azide and mitomycin C) and should be performed in triplicates. A test item that produces a 2.5-fold increase in revertant in comparison to the spontaneous revertant is considered as positive in the AMES test.
2. In-vitro cytogenetic assay: The test should be performed using CHO cells or human lymphocytes. The cells should be exposed at three log concentrations levels (with and without metabolic activation, S9 mix). A reference compound (cyclophosphamide with metabolic activation and Mitomycin C for without metabolic activation) should also be included to record the clastogenic effects over the spontaneous effects. The increase in the number of aberrations in metaphase chromosomes is used to declare positivity.
3. In-vivo micronucleus assay: The preferable species are rodents (preferably mouse) (5animals/sex/dose). The route of exposure of test items should be the same as meant or humans. Mitomycin C or cyclophosphamide should be used as a reference compound. Following the test item administration, the animal may be sacrificed at 24 or 48 h. Femora's bone marrow is harvested, flushed with foetal bovine serum for 20 min, pelleted and smeared on glass slides. The smear should be stained with Giemsa-May-Gruenwald staining. An increase in a number of micronuclei in polychromatic erythrocytes (minimum 1000) cells is considered as a positivity criterion.
4. In-vivo cytogenetic assay: Rat or mouse may be used for the test, but the rat is most preferred (5 animals/sex/dose). The studies should include the positive control (cyclophosphamide) group. Following test item administration, colchicine is intraperitoneally injected at 22 h and 2 h later the animals should be sacrificed. Bone marrow from both the femora should be collected, flushed with

hypotonic saline (20 min), pelleted and re-suspended in Carnoy's solution. The cells should be collected using a Pasteur pipette and dropped (5–10 cm height) on a clean glass slide. An increase in the number of aberrations in metaphase chromosomes (minimum 100) is used for positivity declaration.

1.10.2.8 Carcinogenicity

The test item meant for chronic administration (a minimum of 6 months) in humans has to be evaluated for the carcinogenicity potential. The test is performed in rodent (rat/mouse model) strains with historical evidence, showing low incidences of spontaneous tumours. The test groups comprise vehicle controls and three dose levels (with a minimum 2.5× interval between the doses). The highest dose selected should not affect the life span of the animals by more than 10% of normal expectancy, while the low dose increases the tumour sensitivity. The experimental animals should be treated with the vehicle or test item for a considerable period of life of the animal (24 months for rats and 18 months for mice), which is likely to be the therapeutic exposure period in humans. Observations include mortality and morbidity, periodically and special attention should be paid to ascertain the neoplastic developments and tumour growth. Gross and histopathological changes in organs should be performed.

References

Barile FA (2019) Principles of toxicology testing. CRC Press

Borzelleca JF (2000) Paracelsus: herald of modern toxicology. Toxicol Sci 53:2–4. https://doi.org/10.1093/toxsci/53.1.2

CDSCO (n.d.). https://cdsco.gov.in/opencms/opencms/en/About-us/Introduction/. Accessed 2 Aug 2021

epa-org-chart (n.d.. http://www.orgcharting.com/wp-content/uploads/epa-org-chart-example.png. Accessed 13 Aug 2021

Frank P, Ottoboni M (2011) The dose makes the poison: a plain-language guide to toxicology, 3rd edn, pp 1–xx. https://doi.org/10.1002/9780470918449.fmatter

Gowda C, Sundharam S, Mahadesh J (2014) Mukund Oral-toxicology. J Forensic Dent Sci 6:3–8

Hodgson E (ed) (2010) A textbook of modern toxicology, 4th edn. Wiley, Hoboken, NJ

ICH Guidelines for Pharmaceuticals: Pharmaceutical Guidelines (n.d.). https://www.pharmaguideline.com/2010/10/ich.html. Accessed 21 July 2021

ICH Official web site: ICH (n.d.). https://www.ich.org/. Accessed 26 Mar 2021

ICH Official: Efficacy guidelines (n.d.. https://www.ich.org/page/efficacy-guidelines. Accessed 13 Aug 2021

ICH Official: Multidisciplinary guidelines (n.d.). https://www.ich.org/page/multidisciplinary-guidelines. Accessed 13 Aug 2021

ICH Official: Safety guidelines (n.d.. https://www.ich.org/page/safety-guidelines. Accessed 13 Aug 2021

ICH Topic M 3 (R2) Non-Clinical Safety Studies for the Conduct of Human Clinical Trials and Marketing Authorization for Pharmaceuticals

Imran M, Najmi AK, Rashid MF, Tabrez S, Shah MA (2013) Clinical research regulation in India-history, development, initiatives, challenges and controversies: still long way to go. J Pharm Bioallied Sci 5:2–9. https://doi.org/10.4103/0975-7406.106553

James RC, Roberts SM, Williams PL (2000) General principles of toxicology. In: Principles of toxicology. John Wiley & Sons, Ltd., pp 1–34. https://doi.org/10.1002/0471231800.ch1

Levey M (1966) Medieval Arabic Toxicology: the book on poisons of IBN Wahshiya and its relation to early Indian and Greek Texts. Trans Am Philos Soc 56:1. https://doi.org/10.2307/1005999

OECD - Overview (n.d.) Corporate Finance Institute. https://corporatefinanceinstitute.com/resources/knowledge/economics/oecd/. Accessed 13 Aug 2021

OECD Guidelines for the Testing of Chemicals, Section 1: Physical-Chemical properties (n.d.). https://www.oecd-ilibrary.org/environment/oecd-guidelines-for-the-testing-of-chemicals-section-1-physical-chemical-properties_20745753?_ga=2.261722381.1291410826.1627888750-979487201.1619004932&page=1. Accessed 2 Aug 2021

OECD Guidelines for the Testing of Chemicals, Section 2: Effects on Biotic Systems (n.d.). https://www.oecd-ilibrary.org/environment/oecd-guidelines-for-the-testing-of-chemicals-section-2-effects-on-biotic-systems_20745761?_ga=2.261611149.1291410826.1627888750-979487201.1619004932. Accessed 2 Aug 2021

OECD Guidelines for the Testing of Chemicals, Section 3: Environmental fate and behaviour (n.d.). https://www.oecd-ilibrary.org/environment/oecd-guidelines-for-the-testing-of-chemicals-section-3-degradation-and-accumulation_2074577x?_ga=2.32160318.1291410826.1627888750-979487201.1619004932&page=2. Accessed 2 Aug 2021

OECD Guidelines for the Testing of Chemicals, Section 4: Health Effects (n.d.). https://www.oecd-ilibrary.org/environment/oecd-guidelines-for-the-testing-of-chemicals-section-4-health-effects_20745788?_ga=2.202937745.1291410826.1627888750-979487201.1619004932. Accessed 2 Aug 2021

OECD Guidelines for the Testing of Chemicals, Section 5: Other Test Guidelines (n.d.). https://www.oecd-ilibrary.org/environment/oecd-guidelines-for-the-testing-of-chemicals-section-5-other-test-guidelines_20745796?_ga=2.202937745.1291410826.1627888750-979487201.1619004932. Accessed 2 Aug 2021

OECD Test Guidelines for Chemicals - OECD (n.d.). https://www.oecd.org/chemicalsafety/testing/oecdguidelinesforthetestingofchemicals.htm. Accessed 13 Aug 2021

Pagel 1958 (n.d.) An Introduction To Philosophical Medicine in The Era of The Renaissance (Karger, 1984). Scribd. https://www.scribd.com/doc/244486985/Pagel-Paracelsus-An-Introduction-to-Philosophical-Medicine-in-the-Era-of-the-Renaissance-Karger-1984. Accessed 8 Aug 2021

Pleus RC (2014) Toxicity testing, standards and guidelines. In: Wexler P (ed) Encyclopedia of toxicology, 3rd edn. Academic Press, Oxford, pp 698–702. https://doi.org/10.1016/B978-0-12-386454-3.01042-3

Professionals, I. of M. (US) C. on T. and E.H.I.R. for H, Liverman CT, Ingalls CE, Fulco CE, Kipen HM (1997) The National Library of Medicine's Toxicology and Environmental Health Information Program. In: Toxicology and environmental health information resources: the role of the National Library of Medicine. National Academies Press (US)

Roberts SM, James RC, Williams PL (2014) Principles of toxicology: environmental and industrial applications. Wiley

Schedule Y (Ammended version) (n.d.) CDSCO: Requirements and guidelines for permission to import and / or manufacture of new drugs for sale or to undertake clinical trials

US EPA, O. (2015). Series 870 - Health Effects Test Guidelines. https://www.epa.gov/test-guidelines-pesticides-and-toxic-substances/series-870-health-effects-test-guidelines. Accessed 2 Aug 2021

Wennig R (2009) Back to the roots of modern analytical toxicology: Jean Servais Stas and the Bocarmé murder case. Drug Test Analysis 1:153–155. https://doi.org/10.1002/dta.32

Laboratory Animal Models

2

2.1 Definition

Animal model is defined as 'a living organism in which normative biology or behaviour can be studied, or in which a spontaneous or induced pathological process can be investigated, and in which the phenomenon in one or more respects resembles the same phenomenon in humans or other species of animals' (Chatzigeorgiou et al. 2009).

2.2 Types of Animal Models

The animal models used in biomedical research are mostly classified into the following types: (Schønecker 2014).

(a) **Exploratory model—**the model examines if the impact of a situation (drug/biologics/medical device/surgery) is beneficial or not with regard to a physiological and pathological outcome. It is a simple exploration test.
(b) **Explanatory model—**the model aims to demonstrate the biological mechanisms or mechanism of drug action in disease. It is largely a self-explanatory process.
(c) **Predictive model—**the model is based on some hypothesis or predictive postulation and helps to understand the impact of a treatment, whether it is to cure a disease or to understand the toxicity of a new test compound.

2.3 History

The use of animal models, particularly rodents, in biomedical research has been documented since the sixteenth century. The three main reasons substantiating the importance of the use of animals in biomedical research are as follows:

S. B. Chidambaram et al., *Introduction to Toxicological Screening Methods and Good Laboratory Practice*, https://doi.org/10.1007/978-981-16-6092-4_2

Table 2.1 Genetic Similarity between humans and lab animal models

Animal	Genetic similarity (%)
Chimpanzees	99
Bonobos	99
Gorillas	98
Cat	90
Mice	92
Rat	90
Dog	84
Cow	80

1. In the majority of the conditions, the animals are vulnerable to health issues similar to a human being e.g., diabetes, hypertension, cancer, epilepsy, bacterial and viral infections etc.
2. Animals, particularly mouse models, share 98% DNA similarity to humans. The testing drug's pharmacodynamics, pharmacokinetics and toxicity and safety profiles can be understood very well even before the conduct of a human trial.
3. The short lifespan of animals enables researchers to study the drugs/surgical outcomes quickly.

Similar to humans, the mammalian animal models have complex organ systems that perform distinct physiological functions involving cells, circulatory fluids, hormones, neural networks, growth and thermogenic factors. Thus, the animal model exhibits close phenotypic (anatomical and biochemical) and genotypic features of humans (Table 2.1).

Animal models have fulfilled many medical needs of humans, from the basic understanding of physiology to the development of drugs and vaccines. Some of the major medical outcomes that benefitted humans and animals by the use of animals are listed below:

1. 1898: Discovery of malaria parasite in mosquitoes and birds
2. 1905: First successful human cornea implant in rabbits
3. 1909: First kidney transplant performed in humans in France using animal kidneys
4. 1912: Discovery of Vitamin C in guinea pigs
5. 1914: First successful blood transfusion carried out in dogs, rabbits, guinea pigs
6. 1922: Discovery of insulin using dogs, isolated and tested for use as a treatment in dogs, rabbits
7. 1934: First synthetic anaesthetic discovered in rabbit, rat, monkey
8. 1937: The Rh (rhesus) factor from rhesus monkey into rabbits
9. 1940: Discovery of penicillin as an antibiotic using mice
10. 1955: Development of Polio vaccine in mice and monkey
11. 1955: Advent of the first pacemaker in dogs
12. 1958: First heart transplant in dogs
13. 1960: Discovery of molecules effective as antidepressants studied in rats

14. 1970: Chemotherapy as a treatment for leukaemia in mice
15. 1970: Use of guinea pigs for a testing asthma inhaler
16. 1980: Implementation of magnetic resonance imaging (MRI) as a diagnostic tool in pigs
17. 1987: Zidovudine as the first treatment for HIV tested on mice and monkey
18. 1992: Meningitis vaccine discovered using mice
19. 1994: Aromatase inhibitors for breast cancer approved after testing in mice
20. 2000: Development of vaccines against cervical cancer in rabbits
21. 2000: Anticoagulants were made from animal (goat) proteins
22. 2002: Monoclonal antibodies against rheumatoid arthritis from mice
23. 2006: Advances in stem cell research using rats and mice
24. 2008: First preventive vaccine for cervical cancer using mice
25. 2010: Stem cells were used to repair heart tissues in zebrafish
26. 2011: Cancer treatment using immunotherapy was approved using mice
27. 2016: Discovery of Ebola vaccine from hamster, macaques and rhesus monkey

2.4 Future of Animal Study

Advances in biomedical research technologies paved way for the development of techniques like

- In vitro—In vitro (Latin for within the glass) refers to the technique of performing a given procedure in a controlled environment outside of a living organism.
- In situ—Latin word for 'in site' refers to the experiments where the phenomenon is examined exactly in the place that it occurs, e.g., examining a cell within a whole organ intact and under perfusion system
- Computer models—Computational system biology targets in developing algorithms, tools for visualisation, communications and using data structures to simulate the biological system.
- Organs-on-a-chip (OOC) is a 3D physiological organ biometric system that is built on a microfluid chip, which reflects the structural and functional properties of human tissues by simulating the organ in terms of tissue contact and mechanical stimulation.
- Ex vivo—refers to experiments performed using tissue isolated from an organism in an external environment
- The above-mentioned techniques helped to replace, refine and decrease the number of animals used in biomedical research. However, the development of artificial systems that substitute complex functioning of a living system with pulmonary, cardiovascular and neuronal structures and the complete replacement on the use of animals in biomedical research might be hard.

2.5 Animal Ethics

The historical debate on the use of animals in biomedical research is far from being settled. The developments of modern science and animal welfare awareness have led to the growth of many animal welfare bodies and regulatory frameworks on the use of animals in research. In 1835, the British parliament passed the first Cruelty to Animals Act in 1866. Henry Bergh, a US citizen, established the American Society for the Prevention of Cruelty to Animals. In 1959, William Russell, a UK-based zoologist, described the 3R's principle (Reduction—reduction is the use of a lesser number of animals in a protocol; Refinement—avoiding or minimising pain, use of non-invasive procedures, enriched housing; Replacement—substitution with cell cultures, organs, cellular fractions, microorganism, or lower phylogenetics) (Andersen and Winter 2019). Since the eighteenth century, many animal welfare bodies are developed supporting animal rights in varied ways. People for the Ethical Treatment of Animals (PETA), one of the most popular animal welfare organizations, was established by Ingrid Newkirk, a UK-born animal rights activist. PETA slogan is 'Animals are not ours to experiment on, eat, wear, use for entertainment, or abuse in any other way.'

2.6 India

In India, the animal welfare board was established in 1962, under section 4 of the Prevention of Cruelty to Animals Act, 1960. Committee for the Purpose of Control and Supervision of Experiments on Animals (CPCSEA) was established as a statutory body in terms of section 17 (3) of the Prevention of Cruelty to Animals Act, which was revived in 1988 under the chairmanship of Smt. Menaka Gandhi. Academic institutions and industries involved in research on animals are mandatorily required to register under CPCSEA and comply with rules and regulations. CPCSEA functions with the support of volunteers from academics and industries and liaison with the laboratories involved in research on animal models. At present, the CPCSEA office is located at Krishi Bhawan, New Delhi.

2.7 Functions of CPCSEA

1. To inspect and issue new licenses and renewals of establishments conducting breeding and research on animals.
2. To provide ethical approval for conducting the experiments on large animals (cats, dogs, cattle, goats, sheep, nonhuman primates, and so forth).
3. To recommend legal action against violation of CPCSEA rules and regulations.
4. To select and appoint volunteers as its nominees in the Institutional Animal Ethics Committees of registered establishments.

2.8 Institutional Animals Ethics Committee (IAEC)

The Institutional Animals ethics committee (IAEC) is an eight-member committee that should be formed in an institution, which involves the use of animals for research and breeding activities. The duration of the appointment of IAEC is for a period of 5 years and the respective establishment should reconstitute the IAEC at the time of renewal of license. Primarily, there are four members from the host institution i.e., chairperson, member secretary, scientist-in-charge of animal facility and veterinarian. Also, CPCSEA nominee and link nominee, a socially aware member and scientist(s) from other disciplines/institutions shall be a part of the IAEC.

2.9 Composition of IAEC

Institutional Animals Ethics committee comprises eight members viz:

1. A biological scientist
2. Two scientists from different biological disciplines
3. A veterinarian
4. Scientist-in charge of animal facility of the respective establishment
5. An external scientist
6. A non-scientific socially aware member and
7. A CPCSEA nominee

An expert may be invited while reviewing special projects using hazardous agents such as radioactive substances and toxic microorganisms. The institution nominates the Chairperson and member secretary. CPCSEA nominates members for Serial numbers external scientists, socially—aware members, and CPCSEA nominees.

2.10 Functions of IAEC

1. All research proposals using small animal models, such as rats, mice, rabbits or guinea pigs etc. should be evaluated and approved by the IAEC during the meeting.
2. IAEC cannot approve research projects on large animals; it further recommends the case to CPCSEA.
3. To visit the animal facility where the experiments are performed and to ensure that the respective establishment complies with the CPCSEA rule and regulations ("Compendium of CPCSEA, 2018.pdf," n.d.)

2.11 Laboratory Animals

The most commonly used small animal models are rat, mouse, guinea pig, hamster, gerbils, and rabbit and the large animal models include dog, sheep, goat, pig, and horse in biomedical research. Non-human primate models like baboon, macaque and chimpanzee are used for the higher levels of medical investigations.

Terrestrial invertebrates such as nematodes, species and insects from freshwater, marine life organism crustaceans, planarians, molluscs, etc. are given importance for preliminary investigations based on the suitability and research needs. Specifically, *Drosophila melanogaster*, *Danio rerio* and *Caenorhabditis elegans* are widely used (Table 2.2).

2.11.1 Inbred Strain

Inbred strains are developed by pairing siblings for minimum of 20 or more generations. Inbred strains have high isogeneity and homozygosity that establish phenotypic uniformity and serve as powerful animal models in biomedical research.

2.11.2 F1 Hybrid

F1 hybrids are the first-generation cross of two inbred strains. Like inbred strains, F1 hybrids also show a high degree of isogeneity. However, they show heterogenicity at the loci where the parental strains differ. The heterozygous behaviour promotes a hybrid vigour which can be an aid for certain experiments.

2.11.3 Outbred Strain

Outbred strains are generated by breeding random colonies and maintained for at least four generations. They are regarded as non-isogenic breeds. The genetic compositions are homozygous at very few loci and show significant differences between colonies.

Table 2.2 Taxonomy of routinely used small animal models

	Genus classification				
Animal	Class	Order	Family	Genus	Species
Mouse	Mammalia	Rodentia	Muridae	Mus	
Rat				*Rattus*	*Norvegicus*
Guinea pig			Caviidae	Cavia	Porcellus
Rabbit		Lagomorpha	Leporidae		

2.11.4 Stock Vs. Strain

Stock refers to outbred colonies and strains refer to inbred colonies.

2.11.5 Transgenic Animals

Transgenic animals are produced by deliberately inserting a gene into the genetic material of an animal using genetic engineering techniques. Recombinant DNA technology is employed to develop the gene that is intended to have the desirable characters, through the growth and development of the recipient animal. Transgenic animals express high versatility and hence, they have evolved as a mainstay in biomedical research (Bertolini et al. 2016).

2.11.6 Selection of an Animal Model

The selection of a best-suited animal model depends on the following criterions:

1. Ecological impact
2. Ethical implications
3. Cost and ease of availability
4. Phenotypic equivalence
5. Genetic equivalence, as appropriate and meets minimum requisites
6. Suitability of the model for human disease
7. Experimental feasibility to obtain errorless results
8. Translational potential

2.11.7 Experiment Designing

The investigator should prepare an appropriate study design which should encompass the type of animal model, the number of animals per group, pilot studies, randomisation methods, control groups, and statistical methods to interpret the data and to get the best study outcomes (Festing 2003; Festing and Altman 2002; Johnson and Besselsen 2002a).

2.11.8 Disease Models (Schønecker 2014)

(a) **Induced(experimental) models**
The induced or experimental models involve induction or raising a disease condition using chemical, biological or surgical methods in healthy animals that are phylogenetically close to humans. In most cases, the induced model may

or may not have the exact etiological and pathological effects as that of human pathology. E.g. Induction of diabetes in rats using streptozotocin

(b) **Spontaneous (genetic, mutant) models**
The model utilises the naturally occurring genetic mutations in animals that mimic disease conditions in a human. Also, this model displays a closer phenotypic resemblance between the human disease and that of the host animal. However, impairment in one target gene, sometimes causes activation of other gene(s) and alterations in metabolic processes which can be disadvantageous. One of the best examples of spontaneous mutations is the athymic nude mouse model, the first animal model to explain the natural killer cells.

(c) **Genetically modified models**
The genetically modified animal model or the transgenic animal model uses genetically mutated variants that mimic a human disease condition. Mutating or altering a specific gene in the host animal produces a particular disease that has the genetic and biochemical background of human disease. For example, gene knockout and gene over-expression in animals to establish diseases like diabetes, hypertension, obesity, Parkinson's disease, cancer, etc.

(d) **Negative disease models**
Negative models may be defined as the strain or breed of animals that do not develop certain diseases, i.e., they do not react to certain stimuli used to induce disease. However, this model can be useful for exploring the mechanism of resistance to a particular disease. The negative disease model can sometimes also lead to the characterisation of new spontaneous/mutant models as well. For example, negative models include exploring models of infectious disease or carcinogenicity, where the resistance of an animal to a particular disease is mechanistically studied and used for the development of treatment methods for other disease conditions.

(e) **Orphan disease models**
An orphan model is defined as a functional disorder that might arise in non-human species but has not been identified yet in humans. This implies that when a similar condition is recognised in humans, the data obtained in veterinary medicine earlier is then extrapolated to further assess and define the disease in humans. Some examples of orphan models are Marek's disease and bovine spongiform encephalopathy (BSE) papillomatosis.

2.12 Number of Animals per Group

Allocating animals into various groups and sub-groups is crucial for an experiment and it should be done under the guidance of a biostatistician. Power of analysis (80%) should be calculated to allocate animals to various groups. Before deciding the number of animals, the following information should be decided

(a) The estimated difference between groups
(b) Probable standard deviation

(c) Desired power (80%) and level of significance (5%; $p < 0.05$) (Dell et al. 2002; Festing and Altman 2002)

Each animal should be assigned a unique identification number in order to avoid â error (1-power). The â error or the 1-power indicates the chance of obtaining a false negative result, which means the experiment will reject an untrue null hypothesis or a specified treatment effect (Dell et al. 2002).

2.13 Randomisation

Randomisation refers to the allocation of the number(s) of experimental animals to each treatment group in a way that eliminates bias, to ensure the balance between known and unknown confounding factors that affect the study outcome(Altman and Doré 1990; Festing and Altman 2002). It also provides information on the statistical methods to be employed in analysing the data. Frequently used randomisation methods are:

(a) **Simple randomisation**
It is a single sequence of random allotment of animals to each group. It is simple and easy to implement.
(b) **Block randomisation**
It involves the allotment of an equal number of animals in each treatment group, i.e blocks are small, prefixed with groupings, and remain the same throughout the experiment.
(c) **Stratified randomisation**
In this method, an unique covariate is identified and then the animals are randomised to each treatment group, based on the baseline covariate value.
(d) **Covariate adaptive randomisation**
In this randomisation method, a new animal is systematically introduced to a certain treatment group by taking into account the particular variables and previous assignments of animals.

2.14 Control Groups

A plethora of variables such as genetics, environmental condition, experimental condition, human artefacts, etc. potentially influence the study outcomes. Hence, it is mandatory to include the appropriate control group(s) to minimise the experimental artefacts. There are different types of control groups (Johnson and Besselsen 2002b) (Table 2.3):

Table 2.3 Types of control groups

S No	Control group	Remark
1	Negative control	This group remains the same as a normal state of the animal; may be administered with the vehicle used for toxin/drug preparation
2	Positive control	The disease model is raised and administered with the vehicle used for toxin/drug preparation.
3	Sham control	In this group, a procedure is mimicked (usually surgical) and the vehicle used for toxin/drug preparation is administered. The actual procedure will not be performed.
4	Reference control	Disease raised group administered with known drugs
5	Drug control	Animals treated with a known drug, but no disease raised

2.14.1 Pilot Study

A pilot study or series of pilot studies improve(s) the investigator's skills and in addition, it helps to understand the technical requirements such as the number of animals per group, biological sampling and storing and application of suitable statistics. Importantly, the pilot study utilises the minimum number and also helps to design/modify the experimental procedure to be adopted for the main study (Dell et al. 2002; Festing and Altman 2002; Johnson and Besselsen 2002a).

2.15 Physiological, Clinical Parameters of Laboratory Animal Models

A few important physiological and clinical values of laboratory animals are provided, which will help as reference materials for the experimental pharmacology and toxicology studies.

2.15.1 Physiological Parameters

The basic understanding of the laboratory animals' physiological and behavioural data helps in the correlation of the drug and chemical toxicity testing as well as to understand the disease pathology. The data references for healthy animals are collected from a group of the large number of animals. The below-mentioned parameters may vary with age, diet, environment and genetic factors (Table 2.4).

2.15.2 Blood Clinical Chemistry

The blood clinical chemistry of laboratory animals is recommended to be checked during the toxicity and safety studies. The Blood clinical chemistry parameters like

Table 2.4 Laboratory animals' physiological parameters

	Parameters															
	Life span (year)	Weight (g)[a]		Feed intake	Water intake	Breeding age		Estrus cycle (days)	Gestation period (days)	Weaning period	Blood volume (mL) (M&F)[a]		Heart rate (beats/min)	Respiration rate (breaths/min)	Rectal temp (°C)	Urine volume/day (mL/day)
Species		Male	Female	Single animal		Male	Female				TBV	SVSB/Wk				
Mouse	1–2	20–35	20–35	3–6 g/day	3–7 mL/day	6–8 weeks	6–8 weeks	4–5 (polyestrous)	4–5 (polyestrous)	21 days	2	0.075	300–600	90–180	37.5 °C	1–3
Rat	2–3	350–400	180–200	10–20 g/day	20–30 mL/day	10–12 weeks	8–10 weeks	4–5 (polyestrous)	20–22	21 days	20	1	250–500	80–150	37.5 °C	10–15
Guinea pig	4–6	1000–1200	850–900	20–30 g/day	12–15 mL/100 g	11–12 weeks	7–8 weeks	16–18	65–70	7–14 days	–	–	230–300	60–110	39.5 °C	15–75
Rabbit	5–7	4000–5500	4500–5500	75–100 g	80–100 mL/kg body weight	6–7 months	5–6 months	Polyestrous, induced	30–32 days	6–7 weeks	200	10	250–300	35–55	39.5 °C	50–130

[a]Denotes adult; *TBV* total blood volume; *SVSB* safe volume for single bleed; ((PDF) toxicologist's POCKET HANDBOOK | SIFATEC SA - Academia.edu n.d.)

glucose, urea, creatinine, total protein, albumin and cholesterol are studied to evidence the safety profile of the test item. The reference values are normally compiled based on large numbers of animal's data (Table 2.5).

2.15.3 Haematology

Similarly, in terms of haematology, the reference values are also recommended for animal toxicity and safety studies, e.g. white blood cell, red blood cell count and platelet (thrombocyte) count. Haemoglobin concentration is also used in the diagnosis and monitoring of disease (Table 2.6).

2.15.4 Serum Electrolytes

Serum electrolyte concentrations are the net result of intake and excretion (mainly alimentary and renal). Sodium (Na^+) is the major cation in serum, and its concentration is controlled in concert with the regulation of blood volume and plasma osmolality. Potassium (K^+) is the major intracellular cation and is maintained within narrow limits because of its critical role in neuromuscular and cardiac excitability. Serum potassium concentration should be interpreted with consideration of the acid–base status and potential variations in total body potassium concentration status. Chloride (Cl^-) is the major anion in serum. Extracellular chloride concentration is influenced by extracellular concentrations of sodium and bicarbonate (HCO_3^-) and therefore, interpretation of serum chloride concentration requires knowledge of serum sodium concentration and consideration of acid–base status (Tables 2.7, 2.8, 2.9, 2.10, 2.11, and 2.12).

Table 2.5 Laboratory animals' (adult) blood clinical parameters

	Blood Clinical Parameters											
Animal	Glucose (mg/dL)	Cholesterol (mg/dL)	Triglycerides (mg/dL)	Total bilirubin (mg/dL)	SGOT (IU/l)	SGPT (IU/l)	Creatinine (mg/dL)	Urea nitrogen (BUN) (mg/dL)	γ-GT (IU/l)	Total proteins (g/dL)	Alkaline phosphatase (IU/l)	Albumin (g/dL)
Mouse[a]	115–170	85–150	100–173	0.1–0.5	–	22–90	0.3–0.8	12–27	–	4.2–6.2	30–80	2.7–3.8
Rat[a]	100–200	70–140	50–200	0.1–0.5	45–120	10–50	0.3–1.0	12–20	0–3	6.2–8.0	50–150	3.5–4.0
Guinea pig[a]	75–110	25–80	10–70	0.0–0.3	–	–	0.5–0.8	15–30	5–15	4.5–6.3	50–100	2.2–4.0
Rabbit[a]	100–175	20–60	–	0.1–0.5	10–35	25–70	0.8–1.6	12–22	0–10	5.5–7.0	40–120	3.5–4.7

[a](Genzer et al. 2019; (PDF) toxicologist's POCKET HANDBOOK I SIFATEC SA - Academia.edu n.d.)

Table 2.6 Laboratory animals' haematological parameters

	Haematology									
Animal	RBC ($\times 10^6/\mu L$)	WBC ($\times 10^6/\mu L$)	Haemoglobin (g/dL)	Platelet count ($10^3/mm^3$)	Erythrocyte count ($10^6/mm^3$)	Leukocyte count ($10^3/mm^3$)	Lymphocytes ($\times 10^3/\mu L$)	Monocytes ($\times 10^3/\mu L$)	Eosinophils ($\times 10^3/\mu L$)	Basophils ($\times$103/μL)
Mouse[a]	9.0–11.5	1.0–12.0	13.5–17.5	900–1900	8.0–10.4	4.2–13.3	1.0–9.0	0.0–0.3	0.0–0.5	0.0–0.0
Rat[a]	0.2–0.8	–	15.0–17.8	400–870	6.5–9.6	4.0–18	0.82–5.66	0.020.16	0.1–0.15	0.0–0.03
Guinea pig[a]	4.5–6.8	2.0–12.0	10.9–14.4	175–500	–	–	1.5–10.0	0.0–0.5	0.0–0.5	0.0–0.0
Rabbit[a]	3.7–7.5	5.2–16.5	89–155	112–795	4.8–6.7	3.6–7.9	40–58	2.75–5.5	0.5–2.75	0–0.5

[a](PDF) toxicologist's POCKET HANDBOOK I SIFATEC SA - Academia.edu (n.d.)

Table 2.7 Laboratory animals' serum electrolytes contents

	Serum Electrolyte						
Animal	Sodium (mEq/L)	Potassium (mEq/L)	Chloride (mEq/L)	Calcium mg/dL	Magnesium (mg/kg)	Protein, total (g/dL)	Uric acid (mg/kg)
Mouse[a]	147–163	4.0–7.0	110–128	8.2–11.8	–	4.0–6.0	–
Rat[a]	140–155	3.6–5.9	97–115	9.5–12.0	0.20–1.90	5.7–7.6	8.00–12.0
Guinea pig[b]	120–150	4.0–8.0	90–115	5.3–12.0	3.5–4.1	2.0–5.4	–
Rabbit[a]	132–145	4.0–5.0	96–108	11.0–14.0	0.65–4.20	5.5–7.0	4.00–6.00

[a](PDF) toxicologist's POCKET HANDBOOK I SIFATEC SA - Academia.edu (n.d.)
[b]Washington and Van Hoosier (2012)

Table 2.8 Routes of administration of test item and dose-volume

	Route of administration and dose-volume									
	Gavage (mL/kg)		IV (mL/kg)		IP (mL/kg)		SC (mL/kg)		IM (mL/kg/site)	
Animal	(Ideal)	(Limit)	(Ideal)	(Limit)	(Ideal)	(Limit)	(Ideal)	(Limit)	(Ideal)	(Limit)
Mouse[a]	10	20–50	5	15–25	5–10	30–50	1–5	10–20	0.1	0.5–1
Rat[a]		20–50	1–5	10–20	5–10	10–20	1	10–20	0.1–1	1–10
Guinea pig[b]		5	5	5	1–10	1–10	1–5	1–5	0.05	0.05
Rabbit[a]		10–20	1–3	5–10	3	5	0.5	1–2	0.1–0.5	1

[a] (PDF) toxicologist's POCKET HANDBOOK I SIFATEC SA - Academia.edu (n.d.)
[b]IACUC Routes of Administration Guidelines

Table 2.9 Blood collection in laboratory animals (Parasuraman et al. 2010) & NC3Rs

Species	Routes	With recovery	Non-recovery/terminal method	Limit of withdrawal in case of recovery
Mice	Retro-orbital	0.2 mL	0.5 mL	No more than 1% of the animal's body weight in one collection or over a 24-h period
	Tail vein	50 μl–0.2 ml	–	
	Saphenous vein	0.15 mL/day	–	
	Cardiac puncture	–	1 mL	
	Abdominal/thoracic blood vessel	–	1 mL	
Rat	Tail vein	0.1–2 mL	–	
	Jugular vein	0.1–2 mL	–	
	Saphenous vein	0.2 mL/day	–	
	Retro-orbital	0.2 mL	0.5	
	Abdominal/thoracic blood vessel	–	Up to 1 mL	
	Decapitation	–	10 mL	
Rabbit	Marginal ear vein/artery	Up to 0.5–10 mL	–	
	Cardiac puncture	–	60–200 mL	
Guinea pig	Decapitation	–	Up to 10 mL	
	Abdominal/thoracic blood vessel	–	Up to 15 mL	
	Cardiac puncture		1–25 mL	
	Blood vessel cannulation	0.1–0.5 mL	–	
Hamster	Saphenous vein	0.15/day	–	
	Retro-orbital	0.1–0.5 mL	–	
	Cardiac puncture	0.5 mL	Up to 5 mL	
Dog	Cephalic vein	2–5 mL	–	
	Jugular vein	2–20 mL	–	

Table 2.10 Commonly recommended anaesthetic agents for laboratory animal experiments as per CPCSEA, Govt of India ("Compendium of CPCSEA, 2018.pdf," n.d.)

Species	Short anaesthesia	Medium anaesthesia	Long anaesthesia
Mice	Isoflurane (inhalation)	Xylazine + ketamine (5 mg/kg + 100 mg/kg i.m.)	Xylazine + ketamine (16 mg/kg + 60 mg/kg i.m./i.p.) Or Urethane (1200 mg/kg i.p.)
Rat	Halothane (inhalation)		
Guinea pig	Isoflurane (inhalation)	Xylazine + ketamine (2 mg/kg + 80 mg/kg i.m.)	Xylazine + ketamine (4 mg/kg + 100 mg/kg i.m.)
Rabbits	Isoflurane (inhalation)	Xylazine + ketamine (5 mg/kg + 15 – 30 mg/kg i.m.)	Xylazine + ketamine (5 mg/kg + 100 mg/kg i.m.)

Table 2.11 Commonly recommended anaesthetic agents for laboratory animal experiments as per CPSCSEA ("Compendium of CPCSEA, 2018.pdf," n.d.)

Drugs (mg/kg)	Mouse	Rat	Hamster	Guinea pig	Rabbit	Cat	Dog	Primate
Ketamine HCL	22–24 i/m	–	22–24 i/m	22–24 i/m	22–24 i/m	30 i/m	30 i/m	15–40 i/m
Pentobarbitone sodium	35 i/v	25 i/v	35 i/v	30 i/v	30 i/v	25 i/v	20–30 i/v	35 i/v
	50 i/p	50 i/p	–	40 i/p	40 i/p	–	–	–
Thiopentone sodium	25 i/v	20 i/v	20 i/v	20 i/v	20 i/v	25 i/v	25 i/v	25 i/v
	50 i/p	40 i/p	40 i/p	55 i/p	–	–	–	60 i/p
Urethane	–	0.75 i/p	–	1.5 i/p	1.0 i/p	1.25 i/v	1.00 i/v	1.00 i/v
	–	–	–	–	1.0 i/v	1.50 i/p	–	–

Table 2.12 Commonly recommended euthanasia procedures in laboratory animals as per CPCSEA, Govt of India ("Compendium of CPCSEA, 2018.pdf," n.d.)

General method	Subclassification	Mouse	Rat	Hamster	Guinea pig	Rabbit	Cat	Dog	Primate
Physical methods	Exsanguination	Not acceptable	Acceptable	Acceptable	Acceptable	Acceptable	Acceptable	Not acceptable	Not acceptable
	Decapitation (for analysis of stress)	Acceptable	Acceptable	Acceptable	Not acceptable	Not acceptable	Not acceptable	Not acceptable	Not acceptable
	Cervical dislocation	Acceptable	Acceptable	Acceptable	Not acceptable	Not acceptable	Not acceptable	Not acceptable	Not acceptable
Inhalation of gases	Carbon monoxide	Acceptable	Acceptable	Acceptable	Acceptable	Acceptable	Acceptable	Acceptable	Acceptable
	Carbon dioxide	Acceptable	Acceptable	Acceptable	Acceptable	Acceptable	Acceptable	Not acceptable	Not acceptable
	Carbon dioxide + chloroform	Acceptable	Acceptable	Acceptable	Acceptable	Acceptable	Acceptable	Not acceptable	Not acceptable
	Halothane	Acceptable	Acceptable	Acceptable	Acceptable	Acceptable	Acceptable	Acceptable	Acceptable
Drug administration	Barbiturate overdose (route)	Acceptable (IP)	Acceptable (IP)	Acceptable (IP)	Acceptable (IP)	Acceptable (IV, IP)	Acceptable (IV, IP)	Acceptable (IV, IP)	Acceptable (IV, IP)
	Chloral hydrate overdose (route)	Not acceptable	Not acceptable	Not acceptable	Not acceptable	Acceptable (IV)	Acceptable (IV)	Acceptable (IV)	Acceptable (IV)
	Ketamine overdose (route)	Acceptable (IM/IP)	Acceptable (IM/IP)	Acceptable (IM/IP)	Acceptable (IM/IP)	Acceptable (IM/IV)	Acceptable (IM/IV)	Acceptable (IM/IV)	Acceptable (IM/IV)
	Sodium pentothol [Overdose (route)]	IP	IP	IP	IP	IV	IV	IV	IV

NA not accepted; *A* accepted

References

(PDF) toxicologist's POCKET HANDBOOK I SIFATEC SA - Academia.edu (n.d.). https://www.academia.edu/36341584/toxicologists_POCKET_HANDBOOK. Accessed 20 July 2021

Altman DG, Doré CJ (1990) Randomisation and baseline comparisons in clinical trials. Lancet 335: 149–153. https://doi.org/10.1016/0140-6736(90)90014-v

Andersen ML, Winter LMF (2019) Animal models in biological and biomedical research—experimental and ethical concerns. An Acad Bras Ciênc 91:e20170238. https://doi.org/10.1590/0001-3765201720170238

Bertolini LR, Meade H, Lazzarotto CR, Martins LT, Tavares KC, Bertolini M, Murray JD (2016) The transgenic animal platform for biopharmaceutical production. Transgenic Res 25:329–343. https://doi.org/10.1007/s11248-016-9933-9

Chatzigeorgiou A, Halapas A, Kalafatakis K, Kamper E (2009) The use of animal models in the study of diabetes mellitus. In Vivo 23:245–258

Compendium of CPCSEA 2018." (n.d.)

Dell RB, Holleran S, Ramakrishnan R (2002) Sample size determination. ILAR J 43:207–213

Festing MFW (2003) Principles: the need for better experimental design. Trends Pharmacol Sci 24: 341–345. https://doi.org/10.1016/S0165-6147(03)00159-7

Festing MFW, Altman DG (2002) Guidelines for the design and statistical analysis of experiments using laboratory animals. ILAR J 43:244–258. https://doi.org/10.1093/ilar.43.4.244

Genzer SC, Huynh T, Coleman-McCray JD, Harmon JR, Welch SR, Spengler JR (2019) Hematology and clinical chemistry reference intervals for inbred strain 13/N Guinea pigs (Cavia porcellus). J Am Assoc Lab Anim Sci 58:293–303. https://doi.org/10.30802/AALAS-JAALAS-18-000118

Johnson PD, Besselsen DG (2002a) Practical aspects of experimental design in animal research. ILAR J 43:202–206. https://doi.org/10.1093/ilar.43.4.202

Johnson PD, Besselsen DG (2002b) Practical aspects of experimental design in animal research. ILAR J 43:202–206. https://doi.org/10.1093/ilar.43.4.202

Parasuraman S, Raveendran R, Kesavan R (2010) Blood sample collection in small laboratory animals. J Pharmacol Pharmacother 1:87–93. https://doi.org/10.4103/0976-500X.72350

Schønecker B (2014) Student's guide to animal models, 2nd edn

Washington IM, Van Hoosier G (2012) Clinical biochemistry and hematology. In: The laboratory rabbit, Guinea pig, hamster, and other rodents, pp 57–116. https://doi.org/10.1016/B978-0-12-380920-9.00003-1

Toxicological Screening Methods

3

3.1 Acute Oral Toxicity

3.1.1 Purpose

- Acute oral toxicity provides information on the health hazards likely to occur following the single-dose administration of a test item through the oral route
- The test also helps to categorize the hazard class of the test item, as per the Globally Harmonized System (GHS) of classification and Labelling of chemicals
- Acute toxicity study provides information for selecting doses for long-term toxicity studies (28/90-day repeated studies)

3.1.2 Definition

"Acute oral toxicity refers to those adverse effects occurring in oral administration of a single dose of a substance, or multiple doses given within 24 hours" (Test No. 423 n.d.)

3.1.3 Principle

The acute oral toxicity test is intended to provide information on the likely health hazards that appear upon a high-dose exposure to a test item (Research 2020). The test item is administered once at one of the pre-defined doses (5/50/300/2000 mg/kg) in the animals (preferably female rats/mice) in step I. About 48 or 72 h after the step-I dosing, the presence or absence of toxic clinical signs, including the number of deaths, helps to decide to proceed for additional testing with the same dose or with an increase or decrease of the dose for step II. Thus, the procedure helps to classify the test item by fixed LD50 cut-off values (Eaton and Gilbert 2013).

S. B. Chidambaram et al., *Introduction to Toxicological Screening Methods and Good Laboratory Practice*, https://doi.org/10.1007/978-981-16-6092-4_3

3.1.4 Procedure

Species	:	Rodents (preferably rats)
Sex	:	Female (non-pregnant and nulliparous) (where the use of male is mandatory, it should be justified appropriately)
Age (at start of dosing)	:	8–12 weeks old
Acclimatization	:	Minimum 5 days before dosing
Dose	:	The starting dose is chosen, which is possible to produce mortality in a few of the dosed animals. The highest dose of 2000 mg/kg, b.wt is selected only if any prior information suggests its safety. Otherwise, 300 mg/kg, b.wt is selected as the starting dose

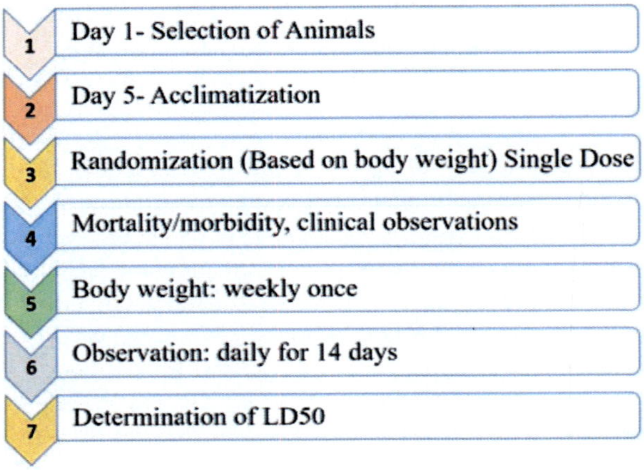

Study state experiment:

1. At the end of acclimatization period, the animals are fasted overnight (~16–18 h in case of rats), but water is provided.
2. The experiment is performed in two steps viz. steps I and II with a minimum 48/72 h interval between the steps. Three animals are used in each step.
3. The bodyweight of the overnight fasted animal is measured prior to dosing.
4. The animal is administered with one of the pre-defined doses as per the flow charts given in figure (Figs. 3.1 and 3.2) and water is withdrawn for the next 4 h thereafter.
5. Following the dosing, the animal is intensively monitored for toxic clinical signs for the next 4 h (preferably at 30 min, 1, 2 and 4 h). After 4 h of intensive monitoring, food and water are provided to the animals.
6. Animals are observed for below mentioned clinical signs.

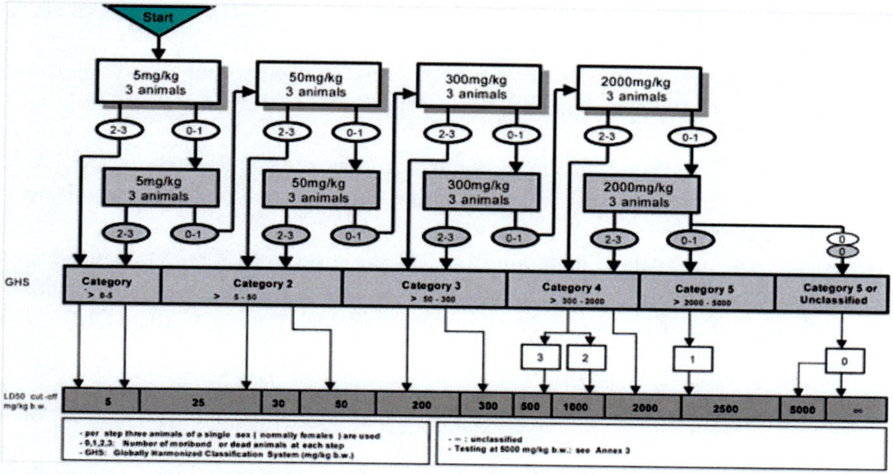

Fig. 3.1 Flow chart for the testing procedure with starting dose 5 mg/kg as per OECD Guideline (Test No. 423 n.d.)

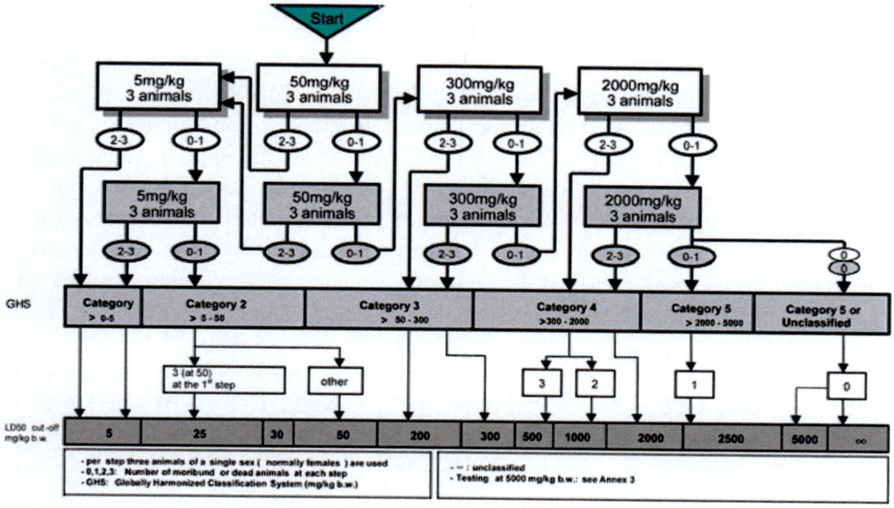

Fig. 3.2 Flow chart for the testing procedure with starting dose 50 mg/kg as per OECD Guideline (Test No. 423 n.d.)

Home cage observation: Moribund, convulsion, tremor, ataxia, hunched posture, kyphosis, lordosis, vocalization and piloerection

Hand held observation: Exophthalmos, lacrimation, visible mucous membrane changes (pale, cyanosis, icteric), nasal discharge, salivation, bleeding from external orifices and so on.

7. Once in a week, a complete functional observation battery is performed to record the signs of toxicity, if any.
8. The animals are observed at least once daily for mortality or morbidity for a total of 14 days. The bodyweight of the animal is measured once in a week and before the necropsy.
9. Any animal(s) found severe moribund during the experiment period, it is euthanized humanely for animal welfare reasons and recorded accordingly (OECD 2000).
10. At the end of 14 days of monitoring, all the test animals are euthanized and subjected to necropsy. Histopathological examination is necessary only if gross pathological changes are observed in the test animal organ(s) (OECD GLP n.d.).

3.1.5 Interpretation

Based on the number of mortalities recorded in steps I and II, the test item is categorized as per the GHS of classification (GHS 2013). Additional information like bodyweight changes, the time course of appearance of toxic clinical signs and reversibility, necropsy and histopathological (if performed) data are summarized to establish the test item classification.

Annexure III (Contd...)
Following administration of test item at 300 mg/kg in three animals in step I, the animals are observed for 48/72 h for mortality, if there are 0–1 animal dead, the same dose is administered in step II. Then, if there are 0–1 animal died, a further higher dose (2000 mg/kg) is tested sequentially (step I and step II). If two to three animals died in step I, then the testing is started with the next immediate lower dose (5 mg/kg). Thus, based on the number of cumulative mortalities recorded (steps I and II), the test item is categorized as per the GHS classification and Labelling of chemicals (Figs. 3.3 and 3.4).

3.2 Acute Oral Toxicity: Step-Up and Step-Down Procedure

3.2.1 Purpose

The test procedure employs a minimum number of animals to possibly evaluate the acute oral toxicity of a test item and also to determine the LD_{50} and confidence intervals that allow for assessments of toxic signs.

3.2.2 Definition

Step-up and step-down procedure is defined as the use of sequential dose pattern to estimate LD50 and confidence intervals of a test item, using a minimum number of

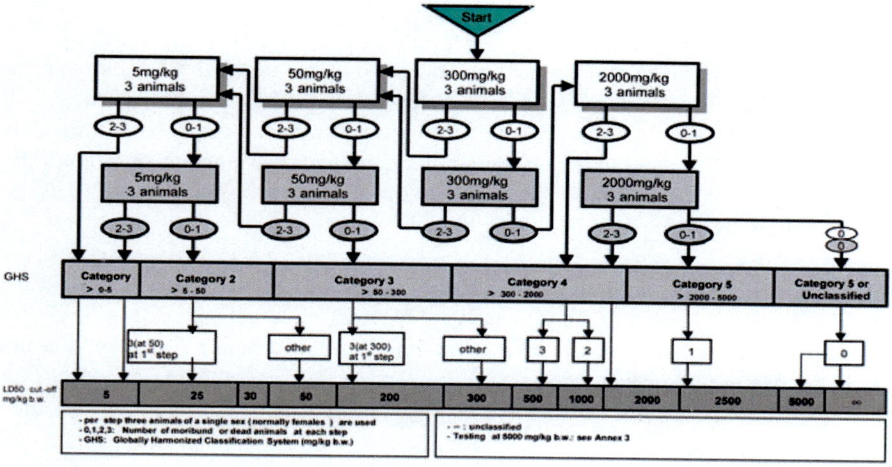

Fig. 3.3 Flow chart for the testing procedure with starting dose 300 mg/kg as per OECD Guideline (Test No. 423 n.d.)

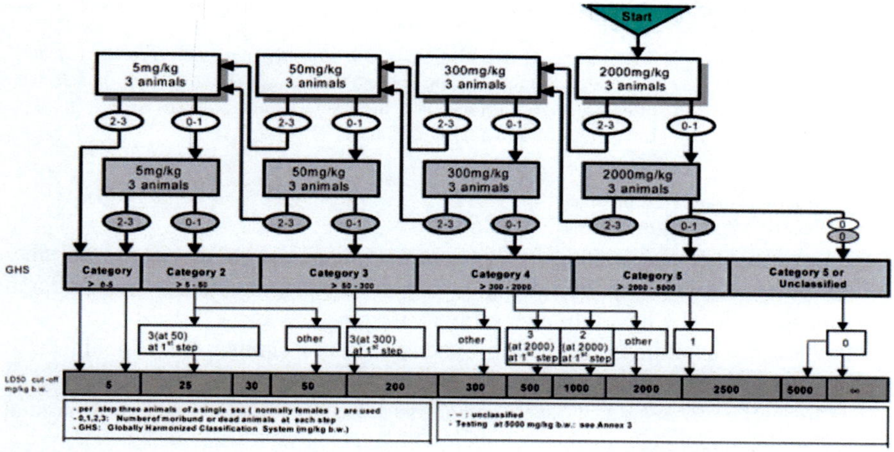

Fig. 3.4 Flow chart for the testing procedure with starting dose 2000 mg/kg as per OECD Guideline (Test No. 423 n.d.)

animals. The dose of the test item is increased (step-up) or decreased (step-down) with a constant factor (3.2 times), based on the mortality as a response indicator (Test No. 425 n.d.).

3.2.3 Principle

The test uses a single animal per dose at a time and is monitored for the next 48 h for mortality if any (starting dose should be a step below the best estimate of the LD50 of the test item). If the animal survives, the next dose for the second animal is increased by 3.2 times. Further, if the animal dies or shows signs of moribund, the dose is decreased by 3.2 times (Research 2020). The dosing stops when one of the below criteria is attained:

1. Three consecutive animals survive at the highest dose tested
2. Five reversals occur in six consecutive animals tested (reversal means a condition where no response is observed at some dose, and a response is observed at the next dose tested, or vice versa)
3. Initial reversal has been followed by at least 4 animals

Procedure		
Species	:	Rodents (preferably rats)
Sex	:	Female (non-pregnant and nulliparous) (where the use of male is mandatory, it should be justified appropriately)
Age (at start of dosing)	:	8–12 weeks old
Acclimatization	:	Minimum 5 days prior to dosing
Dose	:	The starting dose is chosen, which possibly may cause mortality in few of the dosed animals. The highest dose of 2000 mg/kg, b.wt is selected only if any prior information suggests its safety. Otherwise, 175 mg/kg, b.wt is selected as the starting dose.

Study state experiment (Test No. 425 n.d.)

1. An overnight fasted single animal is dosed at a time and monitored for mortality for the next 48 h if any. Following dose administration, the animal is monitored intensively for the next 4 h, preferably at 30 min, 1, 2 and 4 h.
2. If there is no information on the toxicity, the first animal dose should be at 175 mg/kg. If the animal showed morbidity, the next set of animals is treated with a higher dose and if it dies, or appears severe moribund, then the next animal is dosed with a lower dose.
3. Dose sequence to be administered are as follows: 1.75, 5.5, 17.5, 55, 175, 550, 2000 or 1.75, 5.5, 17.5, 55, 175, 550, 1750, 5000 (for specific regulatory requirements)
4. The testing is halted when one of the below-stopping criteria is met:
   (a) three consecutive animals survive at the highest tested dose
   (b) five reversals occur in a six consecutive testing (reversal means a condition where no mortality is observed at some dose and mortality is observed at the next high dose tested, or vice versa)
   (c) A minimum four animals have followed the first reversal
5. Once the stopping criteria are met, LD_{50} is calculated using Self-contained Software for OECD Guideline 425 (Lipnick et al. 1995) (Figs. 3.5 and 3.6).

Stop after animal #7 because 5 reversals in 6 consecutive animals tested (#2-#7).

1	2	3	4	5	6	7	8	9	10	11	12
Step	(I)nclude; (E)xclude	Dose	(X)response (O)non-resp.	Included in nominal *n*	log10 Dose	LD50 =	31.0	LD50 =	12.4	LD50 =	77.6
			OK			Prob. of response	likelihood contribn. (ln *Li*)	Prob. of response	likelihood contribn. (ln *Li*)	Prob. of response	likelihood contribn. (ln *Li*)
1	I	175	X	no	2.2430	0.9335	-0.0688	0.9892	-0.0108	0.7602	-0.2742
2	I	55	X	yes	1.7404	0.6905	-0.3703	0.9020	-0.1031	0.3826	-0.9607
3	I	17.5	O	yes	1.2430	0.3095	-0.3703	0.6174	-0.9607	0.0980	-0.1031
4	I	55	X	yes	1.7404	0.6905	-0.3703	0.9020	-0.1031	0.3826	-0.9607
5	I	17.5	O	yes	1.2430	0.3095	-0.3703	0.6174	-0.9607	0.0980	-0.1031
6	I	55	X	yes	1.7404	0.6905	-0.3703	0.9020	-0.1031	0.3826	-0.9607
7	I	17.5	O	yes	1.2430	0.3095	-0.3703	0.6174	-0.9607	0.0980	-0.1031
8	E				-	-	-	-	-	-	-
9	E				-	-	-	-	-	-	-
10	E				-	-	-	-	-	-	-
11	E				-	-	-	-	-	-	-
12	E				-	-	-	-	-	-	-
13	E				-	-	-	-	-	-	-
14	E				-	-	-	-	-	-	-
15	E				-	-	-	-	-	-	-
Nominal Sample size =				6							
Actual number tested =				7							
Dose-averaging estimator				31.02							
log10 =				1.492							
log-likelihood sums:							-2.2906		-3.2021		-3.4655
likelihoods:							0.1012		0.0407		0.0313
likelihood ratios:									2.4880		3.2378
Individual ratios exceed critical value?				critical=	2.5				FALSE		TRUE
Both ratios exceed critical value?									FALSE		
Calculated maximum likelihood estimate of LD50 =					29.6						

Automated calculation; not relevant to this case.

Final estimate obtained from Maximum Likelihood Calculations

Fig. 3.5 Example of stopping criteria from self-contained software for OECD Guidelines (Test No. 425 n.d.)

Example of stopping criterion (c) from Self-contained Software for OECD Guideline 425

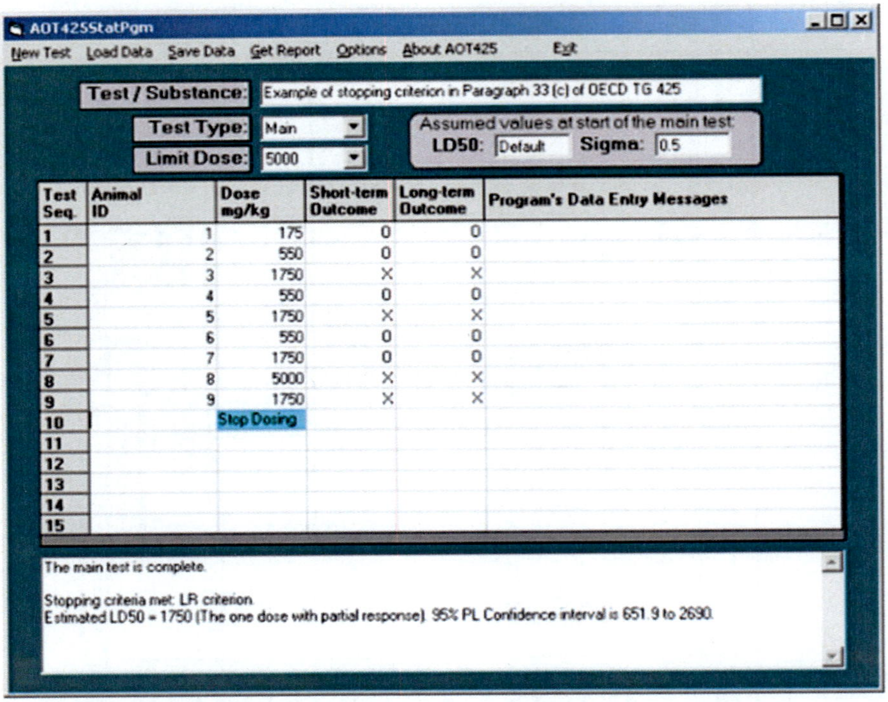

Test Seq.	Animal ID	Dose mg/kg	Short-term Outcome	Long-term Outcome	Program's Data Entry Messages
1	1	175	O	O	
2	2	550	O	O	
3	3	1750	X	X	
4	4	550	O	O	
5	5	1750	X	X	
6	6	550	O	O	
7	7	1750	O	O	
8	8	5000	X	X	
9	9	1750	X	X	
10		Stop Dosing			
11					
12					
13					
14					
15					

Fig. 3.6 Example of stopping criteria (c) from Self-contained software for OECD Guidelines (Test No. 425 n.d.)

6. Animals are observed for mortality or moribund state, twice a day for 14 days, following the administration of the test item. Clinical observations include changes in fur and skin, eyes, mucous membranes, cardiovascular, respiratory, convulsion, tremor, ataxia, vocalization, salivation, lacrimation, abnormal behaviour and so on. The full functional observation battery should be performed once in a week to record the signs of toxicity if any (OECD GLP n.d.).
7. The body weight of the test system should be measured from acclimatization and weekly once thereafter, till necropsy
8. At the end of 14 days of monitoring, all the test animals are euthanized and subjected to necropsy (OECD 2000). Histopathological examination is necessary only if gross pathological changes are observed in the test animal organ(s).

3.2.4 Interpretation

Mortality serves as an index to estimate LD50 and confidence intervals of a test item (GHS 2013).

3.3 Acute Eye Irritation/Corrosion in Rabbit

Eyes are major and complex sensory organs, which serve as the gateways for the entry of various environmental images and colours to the brain through the optic nerve. The eye is broadly divided into two segments viz. anterior segment (conjunctiva aqueous humour, cornea, ciliary body, iris and lens) and posterior segment (sclera, choroid, vitreous humour, retina). Amongst the various routes of drug administration for treating ocular ailments, the topical route is preferred for the anterior and periocular segments while the intravitreal route is mainly used for the posterior segment. The drugs intended for topical administration, that is the anterior application undergoes eye irritation test in rabbit model prior to human use. The chemicals/test items with strong acidic (≤2) and alkaline (≥11.5) pH are not encouraged for in vivo eye irritation testing (Fig. 3.7).

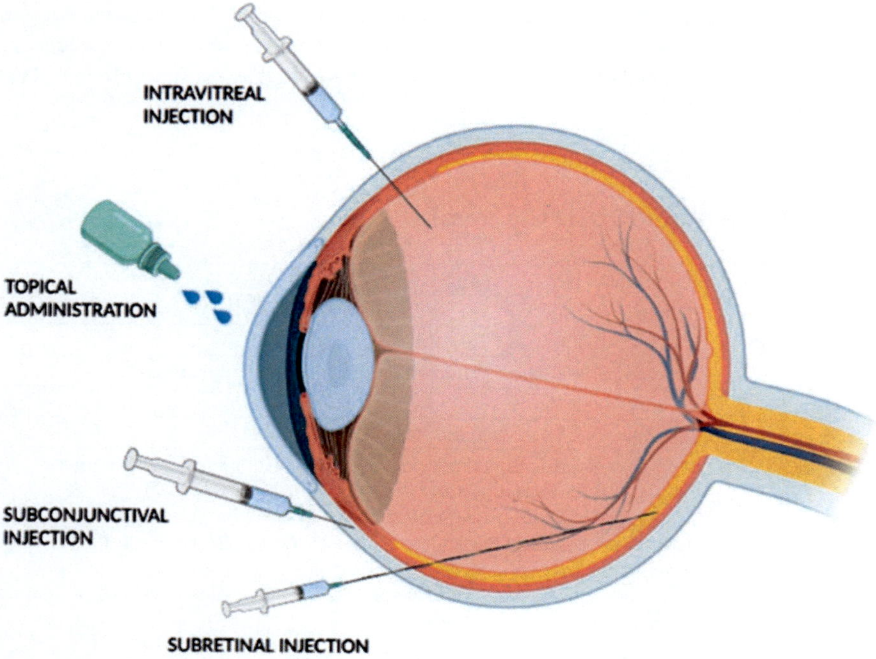

Fig. 3.7 Method of Ocular drug administration at various sites. [Adapted from Tsai et al. (2018) with modifications]

3.3.1 Purpose

Acute Eye Irritation/Corrosion study determines the corrosive and irritant properties and adverse reactions (redness, itching, lesions, etc.) likely to arise following ocular application of a test item

Definitions		
Eye irritant	:	The test item that causes reversible damage to the eye is termed as "Eye Irritant"
Eye corrosive	:	The test item that causes irreversible damage to ocular tissues are known as "Eye Corrosive Agent"
Severe irritant	:	The test item that causes permanent ocular tissue damage or serious vision damage are termed as a severe irritant

3.3.2 Principle

The acute eye irritation/corrosion study involves single-dose administration/instillation of the test item in one eye whilst the contra-lateral eye is considered as a control, in a rabbit model. The intensity of irritation and corrosion caused by the test item is evaluated by grading the score (redness, itching, lesions, etc.) in conjunctiva, cornea, iris and chemosis at different time intervals. The other systemic adverse effects are also recorded to measure the irritant/corrosive nature of the test item effects (Test No. 405 n.d.).

3.3.3 Procedure: (OECD GLP n.d.)

Species	Rabbit (preferred test system)
Strain	Albino
No of animals	3 (1 for initial test and 2 for the confirmatory test)
Sex	Male/female (female should be non-pregnant and nulliparous)
Age	Healthy young adult
Acclimatization	Minimum 5 days prior to the start of the study
Dose level	• Test items with strong acidic ($\leq$2) and alkaline ($\geq$11.5) pH are not encouraged for in vivo testing • Liquid: 0.1 mL to be installed • Paste, solid and particulate substances:0.1 mL/100 mg, test item should be grounded to a fine powder • Aerosols: Test item instillation to the eye in a simple burst of about one second, from a 10 cm distance

Study state experiment:

1. The rabbits with no signs of ocular defects or pre-existing, eye irritation, corneal injury are selected for the study (examined 24 h prior to the study).

2. Rabbits are treated with topical anaesthetic and systemic analgesic before the test item administration. This is to avoid or minimize pain and distress if any of them might arise.
3. Buprenorphine (0.01 mg/kg, s.c; analgesia) is administered 1 h prior to the test item administration. Also, 5 min prior to the test item administration, one or two drops of 0.5% proparacaine hydrochloride or 0.5% tetracaine hydrochloride (ocular anaesthetics) are applied in both eyes. In order to maintain a steady-state analgesic effect, a second dose of buprenorphine (0.01 mg/kg, s.c) and meloxicam (0.5 mg/kg, s.c) are administered 8 h after the test item administration.
4. **"Rescue"**—If an animal shows signs of pain or distress during the study, a "rescue" dose of buprenorphine (0.03 mg/kg, s.c) is given and repeated as required every 8 h following the test item administration.
5. **Testing items**—In the case of liquid, a dose of 0.1 mL is used; in the case of solid, or paste or particulate substance a volume of 0.1 mL or 100 mg is used for application. In the case of aerosols or pump sprays, the test item is extracted from the container and applied. In the case of pressurized aerosols, it is sprayed on the anterior segment of one of the eyes from a distance of 10 cm (Research 2020).
6. **Test item instillation and washing**—The test item is instilled in the conjunctival sac of one of the eyes by gently pulling down the lower lid. After instillation, both the upper and lower lids are held together for a second to prevent the loss of the test item. The contralateral eye serves as the control. After instillation of the test item, the eyes are not washed for at least 24 h unless a serious corrosive/irritant sign is observed. If a solid test item is not absorbed, it is removed by washing with saline or distilled water.
7. **Initial test**—One animal is used in the initial test. If there are signs of corrosiveness and severe irritant is recorded, further studies are not recommended.
8. **Confirmatory test**—If no corrosive reaction is recorded in the initial test, a negative or irritant response is confirmed by using two more animals. To reconfirm a positive result (corrosive and severe irritant) in the initial test, it is recommended that a confirmatory test may be performed in a sequential manner with one animal at a time. If the same effects (corrosive or severe irritant) is observed in the second animal, the study should be terminated
9. **Observations**—Following the application of the test item, the ocular reactions are graded at 1, 24, 48 and 72 h following the standard grading system (Test No. 405 n.d.). If the animal did not develop any ocular reaction, the experiment is terminated at 72 h. If there are mild to moderate irritant reactions, the animals are monitored for the reversibility or irreversibility of reactions till 21 days of the experiment. Fluorescein dye and hand-held slit lamp biomicroscope should be used for the assessment of ulceration or lesions in the eyes.
10. The animals are monitored for weekly bodyweight (in case of 21 days experiment), mortality and morbidity twice a day until the completion of the experiment.

11. Criteria for humanely sacrifice—Following the instillation of the test item, animals developing corneal perforation or ulceration including staphyloma, high-grade corneal opacity (≥4), loss of reflex to light, ulceration in conjunctival membrane and so on, are humanely sacrificed for animal welfare reasons and also these lesions are generally not reversible (OECD 2000).

3.3.4 Grading of Ocular Lesions: (Table 3.1, 3.2, 3.3, and 3.4)

Table 3.1 Corneal injury scoring

Cornea		Grade
Opacity	Degree of density (readings should be taken from the densest area)[a] No ulceration or opacity	0
Scattered or diffuse areas of opacity (other than slight dulling of normal lustre)	Details of iris clearly visible	1
Easily discernible translucent area	Details of iris slightly obscured	2
Nacrousarea	No details of iris visible; the size of pupil barely discernible	3
Opaque cornea	Iris is not discernible through the opacity	4

Maximum possible score: 4

[a]The area of corneal opacity should be noted

Table 3.2 Iris injury scoring

Iris	Grade
Normal	0
Markedly deepened rugae, congestion, swelling, moderate circumcorneal hyperaemia; or injection; iris reactive to light (a sluggish reaction is considered to be an effect	1
Haemorrhage, gross destruction or no reaction to light	2

Maximum possible score: 2

Table 3.3 Conjunctivae injury scoring

Conjunctivae	Grade
Redness (refers to palpebral and bulbar conjunctivae; excluding cornea and iris)	
Normal	0
Some blood vessels are hyperaemic (injected)	1
Diffuse, crimson colour, individual vessels not easily discernible	2
Diffuse beefy red	3

Maximum possible score: 3

Table 3.4 Chemosis grading

Chemosis	Grade
Swelling (refers to lids and/or nictating membranes)	
Normal	0
Some swelling above normal	1
Obvious swelling, with partial eversion of lids	2
Swelling, with lids about half-closed	3
Swelling, with lids more than half-closed	4

Maximum possible score: 4

Table 3.5 GHS Classification criteria for eye irritation under GHS

Category	Description
Category 1	A substance is classified as series Eye damage category 1 (irreversible effect on the eye) when it produces: (a) At least in one tested animal effects on the cornea, iris, conjunctiva that are not expected to reverse or have not fully reversed within an observation period of normally 21 days and /or (b) At least two of three tested animals a positive response of: I. Corneal opacity ≥ 3 and/or II. Iritis >1.5 Calculated as mean scores following grading at 24, 48 and 72 h after instillation of the substance
Category 2A	A substance is classified as Eye irritant category 2A (irritating to eyes) when it produces in at least in two of three tested animals a positive response of: I. Corneal opacity ≥ 1, and/or, II. Iritis ≥ 1, and/or III. Conjunctival redness ≥ 2, and/or IV. Conjunctival edema (Chemosis) ≥ 2 Calculated as mean scores following grading at 24, 48 and 72 h after instillation of the substance, and which fully reverses within an observation period of normally 21 days
Category 2B	An eye irritant is considered mildly irritating to eyes (Category 2B) when the effects listed for category 2A above are fully reversible within 7 days of observation

3.3.5 Interpretation

Based on the ocular grading score, the test item is classified as category 1, category 2A and category 2B (GHS 2013) (Table 3.5).

3.4 Acute Dermal Irritation/Corrosion

Skin is the major organ in the human body which accounts for approximately 10% of the body mass. The skin helps to maintain homeostasis, appreciate sensations (heat/cold, pain, pressure, allergens, etc.) and protect against microbial entry into the body. It is composed of three layers namely the dermis and epidermis hypodermis. The uppermost layer of the epidermis is known as the stratum corneum which performs the barrier functions and prevents the loss of internal components of the body, water

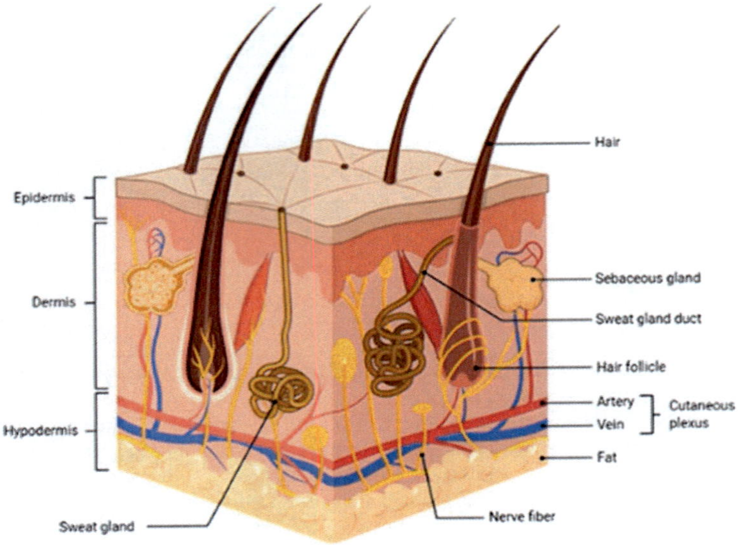

Fig. 3.8 Anatomy of the skin

and so on. The dermis has extensive vascular supply which in turn helps in nutritive supplementation to skin and repair mechanisms involving the immunological components. The other appendages of the skin include hair follicles, sweat glands, sebaceous glands and nails (Fig. 3.8).

Drugs/test items usually permeate through the intact epidermis via skin appendages while the Stratum corneum acts as a permeation limiting factor. In general, chemicals/drugs enter into the body by two pathways

1. Intracellular lipid route, that is between the corneocytes
2. Transcellular route, that is through crossing the corneocytes and the intracellular lipids (Fig. 3.9)

The biophysical characteristic of the skin varies with the anatomical site, age, genetics, geography, lifestyle and so on, which plays a crucial role in drug absorption or sensitivity when applied topically. It is a mandate that the drug meant for topical application needs to be investigated for its skin sensitization and dermal corrosion/irritation potential in animal models prior to human use (Zaid Alkilani et al. 2015).

3.4.1 Purpose

The acute dermal irritation/corrosion study helps to characterize a test item as an irritant/corrosive agent, following a single-dose application in a rabbit model.

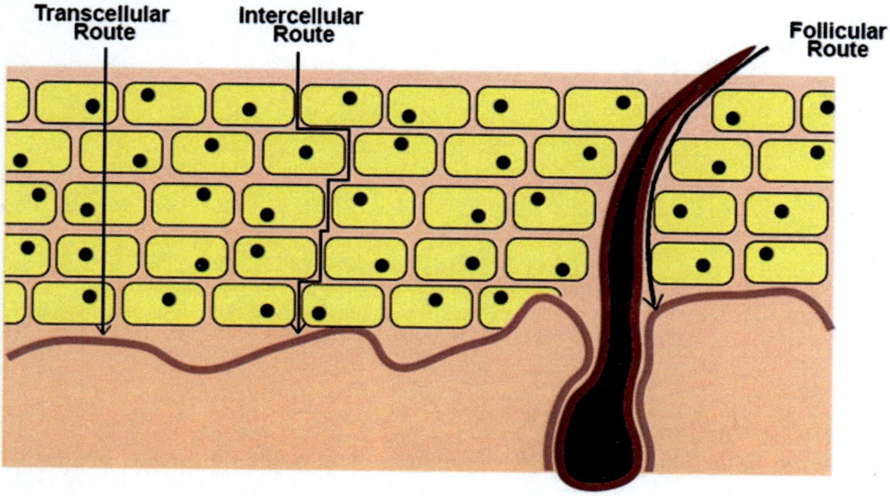

Fig. 3.9 A schematic representation of potential routes of drugs throughout the skin

3.4.2 Definitions

3.4.2.1 Dermal Irritation

The reversible damage caused by acute dermal application of a test item is known as "Dermal irritation"

3.4.2.2 Dermal Corrosion

The localized and irreversible damage to the skin following a single dose application of a test item is termed as "Dermal corrosion". The clinical signs of dermal corrosion include alopecia, bleeding, ulceration, visible necrosis (epidermis to the dermis) and permanent scarring in the experimental animal (GHS n.d.).

3.4.3 Principle

The test item is applied in a single dose (liquid: 500 μL; solid: 500 mg) with uniform and good contact over the skin (fur shaved) of the rabbit for a period of 4 h. At the end of the exposure period, the test item is separated with water/appropriate vehicle carefully, without affecting the skin texture. The degree of irritation/corrosion, if any, is scored and data are interpreted to provide classify the irritant/corrosive nature of the test item (Test No. 404 n.d.).

3.4.4 Procedure: (OECD GLP n.d.)

Species	:	Rabbit
Strain	:	Albino
No of animals	:	3 (1 animal for initial test and 2 animals for confirmatory test)
Sex	:	Male/female (female should be non-pregnant and nulliparous)
Age	:	Healthy young adult
Acclimatization	:	Minimum 5 days prior to the start of the study
Dose level	:	• Test item should not be a strong acid or strong alkali • Liquid: 500 μL should be applied in the undiluted form • Solid:500 mg should be applied as solid powder or paste prepared using water/suitable vehicle

Study state experiment:

1. The fur in the left and right dorsal regions is removed 24 h prior to the test by close clipping with utmost care to prevent skin abrasion. Each side has three different sites for test item application. Each site is used for different time points viz., 3 min, 1 h and 4 h.
2. **Initial test:** A single animal is used in the initial test. The test item is applied on a smaller area (approximately 6 cm^2) as a thin application and loosely covered with a gauge patch and held in place using a non-irritant tape on the site 1. This patch is detached after 3 min. If there is no severe reaction, the test item is administered at site 2 and removed after 1 h and scored for reaction. The third patch is applied at site 3 and allowed to stay in contact for 4 h when no intense reaction is observed at the end of 1 h. At end of 4 h, the skin reactions are scored. Finally, when no corrosive effects are observed after the third patch removal, the animal is monitored for the next 72 h and then the experiment is terminated. If a corrosive effect/distress/severe pain is observed in any of the three stages, the experiment is ended by washing the test item with water. The reversibility of the corrosive effects and other clinical signs are monitored for the next 14 days. Further, histopathological examination is performed, if the corrosive reaction is observed oecd_gd19.pdf (2000). The contralateral side serves as the control/vehicle-treated site (Fig. 3.10).
3. If the test item is irritant in nature, it should be applied on one side and observed for 4 h
4. **Confirmatory test**—When a corrosive effect has not occurred in the initial test, the confirmatory study is conducted using two more animals with one patch and monitored for 4 h. If an irritant response is detected in the pretest, the confirmatory test is conducted either in a sequential manner.
5. All experimental animals used are monitored for erythema and oedema and the scoring is performed at 1, 24, 48 and 72 h. In the case of the initial test, the scoring is performed immediately after the patch is removed.

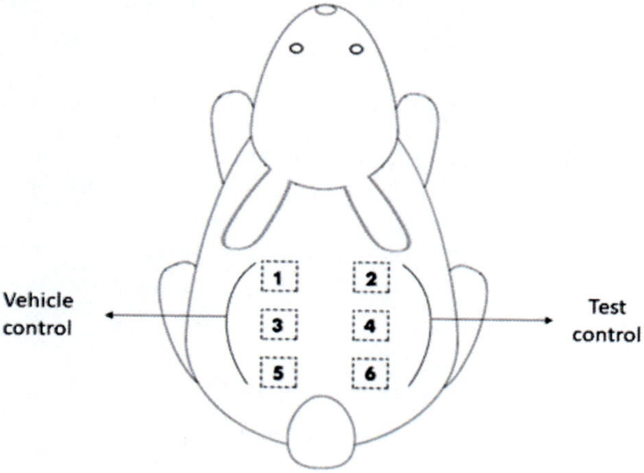

Fig. 3.10 Site of application of test item on animal on contralateral sites

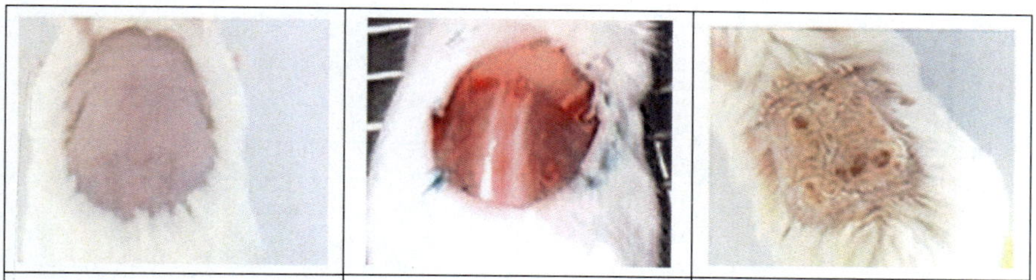

No erythema	Severe erythema (beef redness)	Severe oedema

Fig. 3.11 Severity of erythema formation before and after application of test application. [Adapted from Beshir et al. (2017); He and Tong (2019)]

6. In the case, wherein the irritant or corrosive reaction is not distinguishable the animal(s) is/are observed for the next 14 days, to check the reversibility (Research 2020).
7. The erythema, oedema, dermal reactions are scored as mentioned below (Fig. 3.11):

3.4.5 Grading of Skin Reactions (Test No. 404 n.d.) (Tables 3.6, 3.7, and 3.8)

Table 3.6 Erythema formation

Description	Score formation
No erythema	0
Very slight erythema (barely perceptible)	1
Slight erythema	2
Moderate erythema	3
Severe erythema (beef redness)	4

Table 3.7 Oedema formation

Description	Score assigned
No oedema	0
Very slight oedema	1
Well defined oedema	2
Moderate to severe oedema	3
Severe oedema	4

Table 3.8 Classification of compounds based primary irritation index

Primary irritation index	Classification
0.00	Non irritant
0.01–1.99	Slight irritant
2.00–5.00	Moderate irritant
5.01–8.00	Severe irritant

3.4.6 Interpretation

Based on the erythema and oedema scoring and histopathological examination (if any), the test item is classified as irritant or corrosive (GHS 2013)

3.5 Acute Dermal Toxicity

3.5.1 Purpose

- Acute dermal toxicity study provides information on the health hazard likely to occur following the single-dose application of a test item through the dermal application.
- The study also helps to categorize the hazard class of the test item as per the Globally Harmonized System of classification and Labelling of chemicals.

3.5.2 Definition

Acute dermal toxicity study refers to the evaluation of adverse effects occurring following an uninterrupted exposure to a test item, over a short period of time (24 h or less) (OECD 1987)

3.5.3 Principle

The test item is applied (approximately 10% of the total body surface which is clipped of fur) on the dorsal/flank skin in a single animal. The animal is subsequently monitored for morbidity and mortality for 48 h. If there is no death, the next two animals are exposed to the same dose and monitored for clinical signs of toxicity and death. The test is continued until there is no death in the highest dose tested or death at the lowest dose tested or no more than one death/sign of toxicity in a particular dose and thereby, the hazard class of the test item is categorized (Test No. 402 n.d., p. 402) (Figs. 3.12 and 3.13).

Procedure (OECD GLP n.d.)		
Species	:	Rodents (preferably rats)
Sex	:	Female (non-pregnant and nulliparous)
Age (at start of dosing)	:	8–12 weeks old
Acclimatization	:	Minimum 5 days to dose application
Dose	:	The test starts with a single animal treated with 200 mg/kg bodyweight of the test item applied approximately on 10% of the total body surface area (range-finding study). If there is no death until 48 h, the next two animals should be tested with the same dose (main study). If there is death in the first step, the next lower dose of 50 mg/kg should be tested

Study state experiment

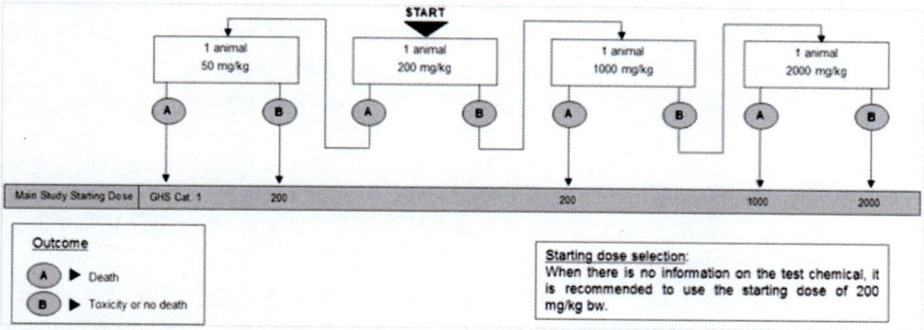

Fig. 3.12 Flow chart for testing procedure; Range finding study starting at 200 mg/kg as per OECD guideline (Test No. 402 n.d., p. 402)

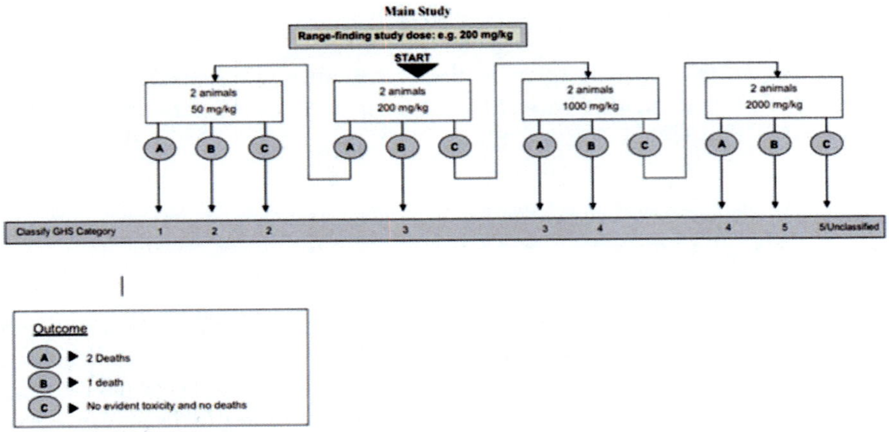

Fig. 3.13 Flow chart for the testing procedure; Main study starting at 200 mg/kg as per OECD guideline (Test No. 402 n.d., p. 402)

- At the end of the acclimatization period and 24 h before the test item application, the fur on the dorsum/flank should be removed and care must be taken to avoid abrasion in the skin. The fur clipped area should be at least 10% of the total body surface (Fig. 3.12).
- If the test item is solid, it should be grounded into fine powder and moistened with a suitable vehicle to apply as a thin film on the clipped area. If the test item is liquid, it should be applied as such (without any dilutions).
- When there is no data on the chemical structure, physicochemical properties, *in vitro* or *in vivo* toxicity, data on structurally related test substances, the study should be started with 200 mg/kg bodyweight of the test item. The test dose should be applied approximately on 10% of the total body surface area (range-finding study). The test item applied site should be covered with gauze dressing, and precaution should be taken to avoid ingestion of the dressing material by the animal. The animal should be kept in an individual cage to avoid the ingestion of dressing material/test item by other animals in the cage. Further, the animal should observe intensively for toxicity signs at 30 min, 1, 2, 4 and 24 h after the test item application and thereafter, twice a day till 14 days.
- The residual test item should be cleaned using water or a suitable vehicle 24 h later.
- If there is no death till 48 h, the next two animals should be tested with the same dose and further dosing should be continued with higher doses of 1000 and 2000 mg/kg (main study). If there is death in the first step, the next lower dose of 50 mg/kg should be tested.
- Animals are observed at least twice a day for mortality and morbidity for a total of 14 days. The bodyweight of the individual animal should be measured before the application of the test item and thereafter, once in a week and before the necropsy (Research 2020).

- At the end of the observation period, all the surviving animals should be euthanized and necropsied to record gross pathological changes, if any oecd_gd19.pdf (2000). If there are any changes in the site of application of the test item or any organ in gross pathology evaluation, it should be taken up for further histopathological assessments.

3.5.4 Interpretation

- Based on the number of mortalities recorded in the range/main study, the hazard class of the test item is determined as per GSH (GHS 2013)
- Additional information like bodyweight changes, the time course of appearance of toxic clinical signs and reversibility, necropsy and histopathological (if performed) data will also help for categorization of the test item).

3.6 Skin Sensitization Test (Guinea-Pig Maximization Test Method)

3.6.1 Purpose

To provide the estimates of relative sensitizing potency and to categorize skin sensitizing agents.

3.6.2 Definition

Skin sensitization is an allergic reaction caused by the disproportionate activation of cell and humoral mediated immune responses, following single or multiple exposures to an allergen/test item. The sensitization reactions may vary from local to systemic effects (Basketter et al. 2008).

3.6.3 Principle

The test is based on the potential of a test item to induce an immunological reaction when administered through intradermal injection and/or epidermal route in Guinea pigs. This period is known as the induction exposure period. After the induction period (minimum 7 days of test item administration, during which time an immunological response may emerge), the test system is again exposed to the test item. This is known as the challenging dose. The degree of skin sensitization in the treated animals is compared with the control animals. In the case of control animals' the test item is administered only during the challenging period but not in the induction period (Test No. 406 n.d., p. 406).

Procedure		
Species	:	Guinea pigs
Sex	:	Male and Female (nulliparous and non-pregnant)
Bodyweight (at dosing)	:	250–400 g
Acclimatization period	:	Minimum 5 days prior to experiment initiation
Supplementation	:	Guinea pigs should be supplemented with Vit-C

3.6.4 Steady-State Experiment

3.6.4.1 Preparation of Test System

The application area should be clipped for fur 24 h before the intradermal injections or epidermal applications of the test item. Care must be taken to avoid skin abrasion or damage.

3.6.4.2 Pre-test

A pre-test should be carried out to find the appropriate test item concentration for intradermal injections/epidermal application, as well as for the challenge dose. A total number of three animals are used for the pre-test. The actual dose for the main study will be obtained based on the pre-test.

3.6.4.3 Main Study

In the main study, the control group consists of 10 animals and the treated group has consisted of 20 animals. The dose used in the induction period and challenge should be systemically well-tolerated and non-irritant, respectively. These dose levels are determined from the pre-test.

3.6.4.4 Preparation of Injection Sites

In the dorsal region of the test system, fur should be clipped off for approximately 6 × 8 cm in the scapular region. About three pairs of intradermal injections (0.1 mL/site) should be given within the margin of a 4 × 6 cm area in the cleared region as follows (Fig. 3.14):

3.6.5 Day—0/Intradermal injection

3.6.5.1 Control Group

Injection site 1: Freund's complete adjuvant (FCA) and physiological saline (1:1 v/v)

Injection site 2: Vehicle (used for test item preparation)

Injection site 3: Mixture of FCA and physiological saline + vehicle [1:1 v/v]

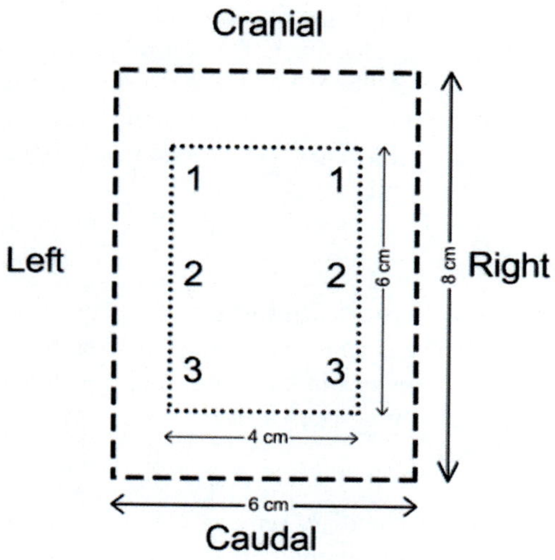

Fig. 3.14 Injection site of the test item on the dorsal surface of Guinea pig

3.6.5.2 Treated Group

Injection site 1: Freund's complete adjuvant (FCA) and physiological saline (1:1 v/v)

Injection site 2: Test item prepared using vehicle (w/v).

Injection site 3: Test item preparation + FCA and physiological saline (1:1 v/v)

A dose volume of 100 μL of the respective preparation is injected in the designated site.

3.6.5.3 Epidermal Application

- On day 5, the fur in the dorsal region is clipped off (approximately 6 × 8 cm) with care to prevent skin damage.
- On day 6, the right side of the scapular region is painted with 0.5 mL of 10% sodium lauryl sulphate (SLS) prepared using Vaseline as base. SLS is applied to cause local irritation.
- On day 7 (i.e. one week after the intradermal injection), a filter (2 × 4 cm dimension) containing the required concentration of test item is placed over the SLS painted right scapular region of the test system. The application is covered with aluminium foil and secured using adhesive tape.
- Similarly, on day 7, in the control group, the vehicle-treated filter paper is applied in the SLS painted right scapular region of the test system and protected using dressing materials.
- On day 8, the dressing material is removed (24 ± 2 h after epidermal application)
- On day 10, the skin reaction following epidermal application is graded according to Magnuson and Kligman grading scale (SCCP 2005)

3.6.5.4 Challenge Dose

- On day 20, the fur in the control and treated animals is removed
- On day 21, the vehicle and test item containing filter paper is applied on the left and right sides of the dorsal region, respectively.
- On day 22, The skin reaction is graded 3 h later following Magnuson and Kligman grading scale
- On day 23, a second observation and grading may be performed, if required

3.6.5.5 Second Challenge

A second challenge is conducted if equivocal results (i.e. an edges incidence of positive reactions in the test group, speedy fading of reactions after the first reading and/or positive reaction in both test and control groups) are obtained after the first challenge.

On day 26, the fur is removed in test and control animals
On day 27, the animals are treated with the vehicle or test items on the right flank in the same manner as described in the first challenge.
On day 28, the dressing material is removed and skin reaction is graded.

3.6.5.6 Observations

Mortality and Clinical signs (Research 2020)

- Mortality should be checked twice daily.
- Clinical signs should be checked once daily from acclimatization till test completion.
- Body weight should be recorded from the day of acclimatization till the completion of test.

3.6.5.7 Skin Reactions

Skin reactions should be observed at the time of reaction during intradermal and epidermal application and challenging periods.

3.6.5.8 Readings and Scoring

The scoring of skin reaction is performed by visual scoring of erythema, oedema and other clinical changes in the skin, as per the Magnusson and Kligman grading scale (Fig. 3.15):

0	=	No visible change
1	=	Discrete or patchy erythema
2	=	Moderate and confluent erythema
3	=	Intense erythema and swelling

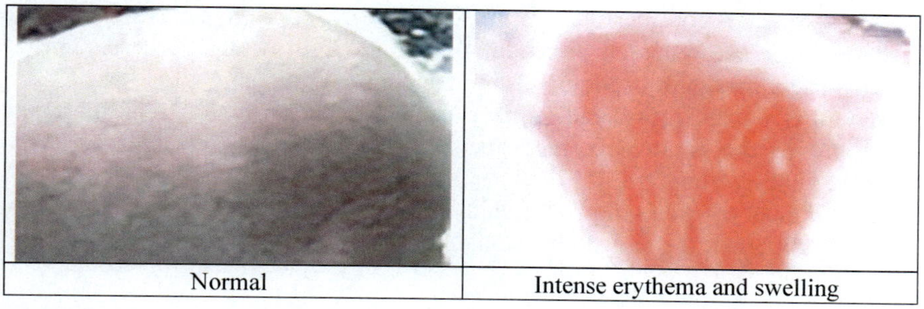

Fig. 3.15 Skin texture/appearance in vehicle and skin sensitizer treated animals [Adapted from Bei-Bei et al. (2016)]

3.6.5.9 Rating of Allergenicity

Based upon the percentage of animals sensitized (24 and 48 h reading), the test item will be assigned to one of the following five grades of allergenic potency, that is ranging from weak to extreme.

Sensitization rate (%)	Grade	Classification
0–8	1	Weak
9–28	2	Mild
29–64	3	Moderate
65–80	4	Strong
81–100	5	Extreme

3.6.5.10 Necropsy

- All surviving animals should be sacrificed at the end of the observation period and gross/macroscopic pathological changes should be recorded.
- Animals which die spontaneously during the observation period should be necropsied as soon as they are found dead and abnormalities should be recorded. The organs showing gross pathological lesions should be collected in 10% neutral buffered formalin (NBF) for histopathological evaluation oecd_gd19.pdf (2000).

3.6.5.11 Histopathology

Histopathological examination should be carried out on those tissues showing gross pathological changes during necropsy (OECD GLP n.d.).

3.6.6 Interpretation

The skin score of the test item treated animals following the challenge application (s) should be compared with the control animals (GHS 2013). The presence of discernible reddening in the reaction site is an indication of an allergic reaction.

- A test system is considered to have no evidence of contact hypersensitivity if the dermal reaction in the test item administered group is identical or less intensive and/or less persistent in comparison to the control group.
- A test system is considered to have contact hypersensitivity if the observed dermal reaction is intensive and/or persistent than any dermal reaction observed in the control group.
- If the dermal reaction score between the control and the test group animals is not clearly different, the test results should be classified as "inconclusive".

3.7 Acute Inhalation Toxicity

3.7.1 Preamble

Chemicals in the form of vapours, gases, mists enter the body through the skin, lungs or digestive system. The safety data of such chemicals is mandatory from the occupational, environmental and human health point of view, of which inhalational toxicity data is one of the major necessities. Chemicals that enter through the inhalational route, intentionally or unintentionally, are generally exhaled or get deposited. Such chemicals enter the upper respiratory tract and diffuse to blood via the lung–blood interface. Depending on the nature of the chemical, mild irritation to severe systemic effects might occur following exposure. Hence, inhalation toxicity data on the single high concentration or repeated exposures is necessary to establish the risk and health hazard potential of the chemicals (Wolff 2018).

3.7.2 Purpose

- Inhalation toxicity studies help us to understand the possible adverse effects that occur from a single high-dose administration of test item through the inhalation route.
- The study also provides information on the median lethal concentration (LC_{50}) and help to categorize the test item based on the toxic/adverse effects as per the GSH

3.7.3 Definition

Acute inhalation toxicity study (nose-only route/whole-body exposure) refers to the evaluation of adverse reactions that occur due to single uninterrupted exposure to a test item (in the form of a gas/vapour/aerosol or a mixture of both) for a period of 4 h, following the traditional LC_{50} protocol (EPA n.d.) (Fig. 3.16).

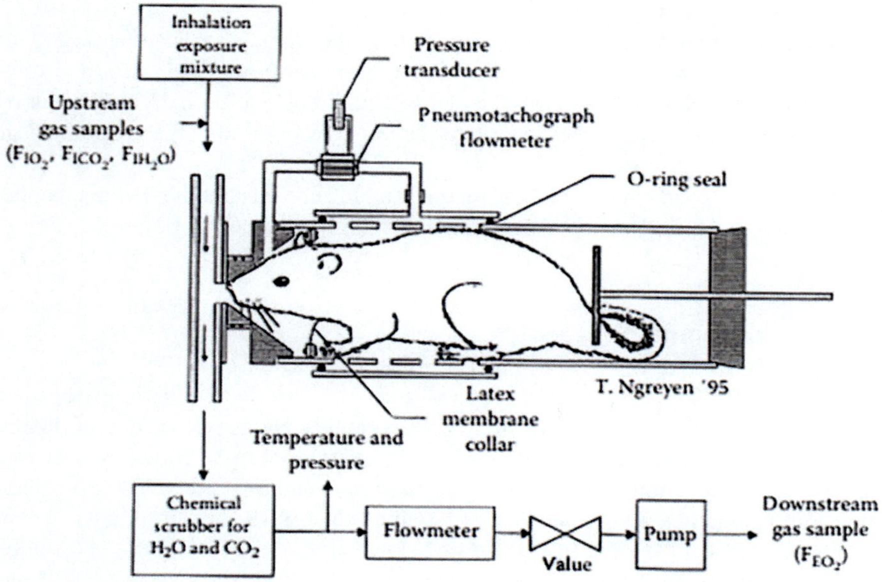

Fig. 3.16 Nose-only exposure apparatus. [Adapted from Salem and Katz (2014)]

3.7.4 Principle

The test is performed by two methods.

Traditional Protocol: The animals are exposed to a limit concentration or three concentration levels (step-wise) of the test item for a period of 4 h

Concentration X Time (C-xT): The animals are exposed to a series of concentrations over multiple time durations.

Procedure		
Species	:	Rat
Sex	:	Males or females or both (nulliparous and non-pregnant)
Age (at start of dosing)	:	8–12 weeks old
Acclimatization	:	Minimum 5 days to dose application
Inhalation chamber	:	Generally fixed based on the nature of the test item. Test system volume (size) should not be >5% of the chamber volume. The concentration of O_2 and CO_2 should be 19% (minimum) and not more than 1%, respectively. The temperature and humidity are maintained at 22 ± 3 °C and 30–70%, respectively. The temperature and humidity should be measured at 1, 2 and 3 h of the exposure period
Test item	:	The test item must be in the form of gas, vapour, aerosol or a mixture of these based on the physicochemical properties. The particle size of the aerosol should be in the range of 1–4 μm MMAD with a GSD of 1.5–3.0. Limit concentrations for gases, vapours and aerosols are 20,000 ppm, 20 mg/L and 5 mg/L, respectively

Study state experiment (Test No. 403 n.d.)

- At the end of the 5 days acclimatization period, the animals are acclimatized to the inhalation chamber for 30 min prior to test item exposure
- Animals exposed whole body to an aerosol should be housed individually during exposure. This prevents the filtering of the test aerosol through the fur of the cage mates.
- A test run is conducted to obtain suitable, stable and constant inhalation chamber exposure conditions without the animals with a limited concentration

3.7.5 Traditional Protocol

Sighting study: This is study is performed to fix the dose for the limit test or main study and also helps to determine the sex vulnerability and toxic potency of the test item. Typically, three males and three females are exposed to 20,000 ppm or 20 mg/L or 5 mg/L or maximum attainable concentration in sighting study. The concentration for the sighting study may also fix based on the QSAR data or toxicity data of similar chemicals, if available. If the tested concentration produced 100% mortality, another sighting study should be conducted with a lower concentration. If the mortality is ≥50% at the tested concentration or at the maximum attainable concentration, the main study is conducted with three lower concentrations.

Limit test: It is performed for the test item which is virtually non-toxic or the toxicity response is greater than the regulatory concentration limits. In limit test, three males and three females are used and exposed to limit concentration (maybe a non-toxic concentration) of the test item. Limit concentrations for aerosols, gases and vapours are 20,000 ppm, 20 mg/L and 5 mg/L, respectively. In the case of aerosol testing, the particle size should be 1–4 μm MMAD with a GSD of 1.5–3.0. If the mortality is ≤50%, no further testing is required.

Main study: The main study is performed with three concentration levels. Typically, five males and five females, per concentration (initially with a lower concentration), are exposed to the test item for a period of 4 h. The clinical signs of animals are monitored every 30 min for 4 h. Exposure to the next higher concentrations will be scheduled until the confidence in the survival of pretested animals is attained.

3.7.6 Concentration × Time (C × T) Protocol

(C × T) the protocol is an alternate model to the traditional protocol in assessing inhalation toxicity of test items. This model generally uses two animals per sex with four concentrations and five time points to assess the inhalation toxicity potential. C × T protocol studies are performed only in the nose-only chamber because whole-body chamber is not practically suited for the technique.

Sighting study: In this study, three animals/sex/concentration are exposed to the test item for a period of 4 h. If the mortality is ≥50% for the tested concentration or at the maximum attainable concentration, the main study is conducted with three lower concentrations

Initial concentration study: In this study, two animals (1 per sex)/group/concentration are exposed to different time points viz., 15, 30, 60, 120 and 240 min. If the mortality is ≤50%, no further testing is required.

Main study: Based on the sighting or initial concentration study, the concentration for the main study is fixed. In the main study, five males and five females are exposed to a fixed concentration for different time points as mentioned above. The mortality is monitored during the exposure and observation (14 days) periods.

- Following exposure to the test item, the onset, duration, severity and reversal (if any) of toxic signs is recorded. Major clinical toxic signs like changes in the skin, mucus membranes, respiratory, circulatory, somatomotor and behaviour patterns should be recorded. The presence of salivation, tremors, convulsions, lethargy and diarrhoea should also be recorded (OECD GLP n.d.).
- During the observation phase for 14 days (after being exposed to the test item), the animals are monitored for morbidity and mortality, at least once daily (Research 2020)
- The bodyweight of the individual animals is recorded prior to the test item exposure, thereafter on days 1, 7 and 14 and before necropsy
- At the end of the observation phase, animals are euthanized oecd_gd19.pdf (2000) and gross pathological examination should be performed with special attention to the respiratory tract. Gross pathology is not mandatory for sighting study.

3.7.7 Interpretation

- Lethality estimates (e.g. LC_{50}, LD_{01}) with 95% confidence limits and slope are determined on the basis of a number of mortality and quantitative risk assessment (the most significant cause of death and the most common way of action) which allows the test item to be classified according to the Globally Harmonized System of Classification and Labelling of Chemicals (GHS) (GHS 2013).

3.8 28/90 Day Repeated Oral Toxicity

Drugs developed for the management of chronic ailments like diabetes, hypertension and so on need to be consumed for a significant proportion of the human lifespan, which in turn means the drug will be available in the systemic circulation continuously. In such cases, the effects of the drug on the function and structure of the visceral organs, the hemodynamic and clinical parameters need to be studied thoroughly. Further, the toxicity is simulated in sub-acute and subchronic toxicological studies, wherein the test item is administered for a significant lifespan of the test system and the relevant investigations are performed to understand the toxic potential (Fig. 3.17).

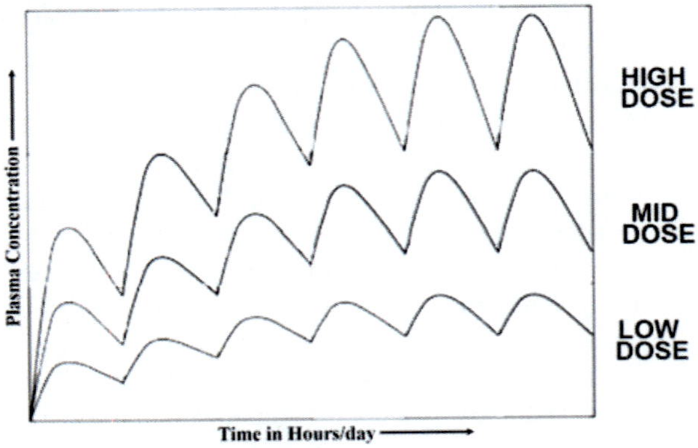

Fig. 3.17 Influence of dose size on plasma concentration-time profile after oral administration of a drug at fixed intervals of time. [Adapted from Design of Dosage Regimens (n.d.) with modifications]

Sub-acute and subchronic toxicological studies are referred as 28 and 90 days repeated dose toxicity studies, respectively. Treatment for a period of 90 days is equivalent to approximately 10% of lifespan exposure in rodents.

3.8.1 Purpose

- To provide information on the cumulative, and target organ toxicity as evidenced from the functional outcomes, clinical biochemistry, gross and histopathological changes following the administration of the test item over a period of time (28/90 days) in the test system
- To establish the "no observed adverse effect level" (NOAEL) and "low observed adverse effect level (LOAEL)" dose level, which helps to decide the dose for further toxicological testing or for phase I clinical trial

3.8.2 Definition

Repeated oral toxicity study refers to the assessment of the toxic potential and the health hazards likely to arise following the oral administration of a test item, over a limited period (28/90 days) of life span of the test system (Hayes et al. 2020).

3.8.3 Principle

The test item is administered orally at three dose levels (per group one dose level) to a healthy test system for a period of 28/90 days. During the dosing period, the test system is observed for clinical signs of toxicity, bodyweight changes, feed and water intake at definite time points. At the end of the dosing period, blood and organs are collected for various investigations. Based on the data viz. bodyweight changes, feed intake, blood clinical chemistry (haematology, electrolyte, clinical parameters), gross and histopathological changes, the no-observed adverse effect level (NOAEL) and low-observed adverse effect level (LOAEL) are determined. Further, the toxic dose range of the test item is also established.

Procedure (OECD GLP n.d.)		
Species	:	Rat is the preferred test system
Strain	:	Common strains like Sprague Dawley, Wistar Albino
Sex	:	Male and female (female should be nulliparous and non-pregnant)
Age at dosing	:	<9-week-old
Acclimatization	:	Minimum 5 prior to the start of the study
Randomization	:	Healthy test systems are selected and randomized based on stratified bodyweight or by random distribution. In any case, the average differences in body weights between the groups should not be statistically significant

Study state experiment (Test No. 408 n.d.)

1. Based on the acute toxicity or limit test, the doses for 28/90-day toxicity are fixed.
2. The highest dose is fixed with an aim to induce toxic signs, but no death or severe suffering in the test system.
3. The descending doses are fixed at two to a fourfolds intervals between the groups, so as to evidence a dose-dependent response, if any (Research 2020).
4. The details of groups and treatment templates are given in Table 3.9. The test systems are dosed daily for 28/90 days, preferably at the same time every day.
5. At the end of the dosing period, the overnight fasted test systems are subjected to blood collection and necropsied (OECD 2000).
6. In the case of satellite groups, the dosing of the test item is stopped after 28/90 days and is observed for delayed or persistence or recovery from the toxic effects for the next 14 days.

Table 3.9 Template of groups and treatment in 28/90 day repeated oral toxicity study

		28 day		90 day	
Groups	Treatment	Male	Female	Male	Female
Control	Vehicle	5	5	10	10
Low dose	Test item	5	5	10	10
Middle dose		5	5	10	10
High dose		5	5	10	10
Satellite–Control	Vehicle	5	5	5	5
Satellite–High dose	Test item	5	5	5	5

3.8.4 Observation

The test systems are monitored during the dosing period and for an additional 14 days without treatment in the case of satellite groups.

Observation		Monitoring period	Specification
Mortality	:	Twice in a day (daily)	The test systems are observed for mortality from the start of acclimatization period and throughout the experiment.
Bodyweight	:	Once in a week	Bodyweight is recorded from the start of acclimatization day until necropsy
Clinical observation	:	Daily (1–2 h after dosing)	Routine clinical signs
	:	Once in a week	Complete functional observation battery
Feed Intake	:	Daily	Recorded from the start of acclimatization day until necropsy; except on the day of fasting
Water intake	:	Daily	When the test item is administered through drinking water
Blood parameters At the end of the treatment period/recovery period	:		**Haematology**: Haemoglobin (HGB), Haematocrit (HCT), total erythrocyte count (RBC), mean corpuscular volume (MCV) mean corpuscular haemoglobin (MCH), platelet count, mean corpuscular haemoglobin concentration (MCHC), total leucocytes (WBC), blood clotting time, reticulocytes, neutrophils, lymphocytes, eosinophils, basophils and monocytes
			Electrolytes: Total calcium, potassium, sodium, chloride, inorganic calcium

Clinical biochemistry	:	Last week of the dosing/ recovery	Carbohydrate metabolism: Glucose Lipid metabolism: Total cholesterol, triglycerides Protein metabolism: Total protein, albumin **Liver function** Hepatocellular: serum glutamyl pyruvate aminotransferase (SGPT) Hepatobiliary: γ-glutamyl transferase (SGOT), total bilirubin, total bile acids Renal function: creatinine and urea

(continued)

Urine analysis (optional)	:	Last week of the dosing/ recovery	Urine volume, appearance, specific gravity, pH, ketone bodies, glucose and blood cells
Ophthalmological examination	:	Last week of the dosing/ recovery	Fluorescein dye and hand-held slit lamp biomicroscope used for assessment of ulceration or lesions in the eyes
Pathology Necropsy is performed in all test systems that survived until the end of the dosing or recovery period. Necropsy is also performed on animals that express undue stress or die during the experiment			
Gross pathology	:	Organs examined in 28 days repeated oral toxicity study	Organs such as external orifices, brain, eyes with the optic nerve, thoracic plug, thyroid and thymus glands, lungs, heart, stomach, spleen, liver, small and large intestines (with Peyer's patches), kidneys, adrenals, urinary bladder, testes, epididymites, uterus with the cervix, ovaries, vagina, peripheral nerve, skeletal muscle, spinal cord, bone with bone marrow, skin with mammary gland are examined for gross pathological changes
	90 day repeated oral toxicity study		Organs such as the pituitary, parathyroid, oesophagus, salivary glands, pancreas and aorta are also examined in addition to the organs mentioned in the 28-day study
Organ weight	:	Immediately after necropsy	Absolute and relative organ weight of brain, heart, liver, paired kidneys, paired adrenals, spleen, paired testes, paired epididymis, male sex glands, uterus with the cervix, paired ovaries, thymus and spleen are measured
Histopathology	:	Organs mentioned in gross pathology are subjected to histopathological examination	Initially, the control and high dose groups are examined. If the organ (s) from the high-dose groups show histopathological changes, organs from the other descending doses are then examined

3.8.5 Data Interpretation

- The mean differences between the control and test item administered groups (low, mid and high dose) are analysed by biostatistical methods.
- The change observed in one parameter is correlated with the associated parameters to rule out the treatment-related toxicity. For example, a decrease in feed intake causes decrease in bodyweight, alterations in plasma glucose, cholesterol, triglycerides and total proteins, ulceration in the stomach and histopathological alterations.
- Based on mortality, morbidity and biologically relevant statistical significance between the control and test item administered groups, the NOAEL and LOAEL are established (GHS 2013).

3.9 28-Day Inhalation Toxicity Study

3.9.1 Purpose

- To establish the No Observed Adverse Effects Concentration (NOAEC), Lowest Observed Adverse Effects Concentration (LOAEC), Maximum Tolerated Concentration (MTC) of a test item exposed through the inhalation route
- This study helps to characterize the chemical toxicity and provide quantitative data on the inhalation limits and risk assessment of the test item

3.9.2 Definition

Sub-acute inhalation toxicity study refers to the assessment of toxic potential and the health hazards likely to arise following the repeated exposure to a test item (gas, vapour and aerosol) at different concentration levels for a period of 28 days (6 h/day for 5 days/week) through inhalation route in male and female rats (Derelanko and Auletta 2014).

3.9.3 Principle

The animals are exposed to a test item for 6 h/day over 28-days duration at more than 3 concentration levels (test groups) and to filtered air (vehicle group) through inhalation routes. The test system is observed for clinical signs of toxicity, bodyweight changes and feed intake at definite time points. The data obtained from the sub-acute inhalation toxicity studies are used for quantitative inhalation risk assessments in terms of target organ toxicity, particularly lungs and to establish No-Observed-Adverse-Effects-Concentration, Low-observed adverse effect concentrations (LOAEC) and reversibility of toxic effects if any, and for the selection of concentration for chronic studies (Test No. 412 n.d.).

Procedure (OECD GLP n.d.)		
Species	:	Rat
Strain	:	Wistar Albino / Sprague Dawley
Sex	:	Male and Female (Nulliparous and non-pregnant)
Age (at the time of dosing)	:	7–9 weeks
Acclimatization	:	5 days prior to start of the experiment. During acclimatization period the animals are adapted to inhalation exposure chamber for 30 min every day. This reduces stress during test item exposure period
Randomization	:	Healthy animals are selected and randomized based on stratified bodyweight or by random distribution. In any case, the mean differences in body weights between the groups should not be statistically significant
Animal husbandry	:	Animals should be housed at a temperature of 22 ± 3 °C, relative humidity between 30 and 70% and 12/12 h light–dark cycle.
Inhalation chamber	:	A dynamic inhalation chamber should be used. Animal size should not exceed 5% of the chamber volume. Concentrations of O_2 should be ≥19% and CO_2 ≤ 1% in the exposure chamber

Test atmosphere: concentration levels		
Nominal concentration	:	The nominal concentration of the test item in the exposure chamber should be determined before the test system is exposed to the test item. It is determined by estimating the concentration of test item in the chamber during the generation period and dividing this value by air flow in the chamber during the same period). Nominal exposure gives information on the generation efficiency of the test system
Analytical concentration	:	The analytical concentration of the test item should be determined before the exposure of the test system to the test item. It is determined by measuring the test item concentration at the animal's breathing zone in the exposure chamber. This is performed to maintain test item concentration homogeneity in the exposure chamber
Limit concentrations	:	According to the GSH system, aerosols have a maximum concentration of 5 mg/L, vapours have a maximum concentration of 20 mg/L and gases have a maximum concentration of 20,000 ppm

Study state experiment

A. **Limit concentration**

- When there is no information on the test item toxicity data, the highest concentration of 5 mg/L for aerosols, 20 mg/L for vapours and 20,000 ppm for gases is used in the limit concentration test.

B. **Pilot study**
   - Prior to the exposure of animals to the test item, a pilot study should be conducted (without animals) to determine formulation efficiency and to ensure the delivery of desired concentration.

C. **Dose range finding study**
   - Prior to the main test, a dose range finding study should be carried out at three or more selected concentration levels with five males and five females per concentration.
   - Animals are exposed to the test item for 6 h for 14 days (5 days/week) and monitored for mortality and clinical signs during the exposure period and in-home cages.
   - At the end of the 14 days of exposure time, the animals are euthanized and subjected to gross pathology. Lung splitting is performed, subjecting right lung for broncho-alveolar lavage fluid (BALF) analysis and left lung for histopathology. BALF analysis includes the estimation of lactate dehydrogenase (LDH), total protein, albumin, alveolar–macrophages, lymphocytes, neutrophils, eosinophils and differential cell counts.
   - If the test item possesses pulmonary deposition and retention (due to poor solubility) properties, then the lung burden measurement should be performed by bioanalytical estimation of the test item. Estimation of the test item in lung-associated lymph node serves as an indicator for translocation. (Translocation is the particle burden of relevant organ)
   - BALF and lung burden measurement should be performed within 24 h following the last exposure to the test item, that is post-exposure observation (PEO). Additional PEOs may be planned to generate information on the pulmonary retention and clearance kinetics of the test item
   - In order to establish robust toxicological information from the dose range findings study, the blood parameters (haematology, serum electrolytes, clinical chemistry) and histopathology are performed
   - These data help to establish the concentrations for the main study. The highest concentration of the test item for the main study should be selected in such way that it does not cause mortality but impart signs of toxicity in animals

D. **Main study**
   - The main study should be conducted at a minimum of three concentration levels with five males and five females per concentration. Groups and treatment template is shown in Table 3.10
   - Animals are exposed to test items using a dynamic inhalation chamber for 6 h/day, 5 days/week for a total period of 28 days.
   - Animals in the control group should be exposed only to filtered air
   - Animals in the satellite groups must be exposed to the test items for 28 days and a free period of 14 days following last exposure to determine the delayed onset or persistence of toxicity.
   - If any animal is found dead/moribund during the exposure period it should be subjected to detailed necropsy and the death time should be recorded oecd_gd19.pdf (2000).

Table 3.10 Experiment design and treatment

Groups	Treatment	Sex	
		Male	Female
Control	Vehicle	5	5
Low concentration	Test item	5	5
Middle concentration		5	5
High concentration		5	5
Satellite–Control	Vehicle	5	5
Satellite–High concentration	Test item	5	5

3.9.4 Observations

Test systems should be monitored during the test item exposure period, and for an additional 14 days without treatment in case of the satellite groups (Research 2020).

Observation		Monitoring period	Specification
Mortality	:	Twice in a day (daily)	Test systems should be monitored for mortality from the start of the acclimatization period and throughout the experiment
Body weight	:	Once in a week	Body weight is recorded from the start of acclimatization day till necropsy
Clinical observation	:	Daily (minimum 1–2 h after dosing)	Routine clinical signs
		Weekly once	Complete functional observation battery
Feed Intake	:	Daily	Recorded from the start of acclimatization day till necropsy; except on the day of fasting
Blood parameters	:	At the completion of treatment period/recovery period	**Haematology**: Haemoglobin (Hb), haematocrit (HCT), total erythrocyte count (RBC), mean corpuscular volume (MCV), mean corpuscular haemoglobin (MCH), mean corpuscular haemoglobin concentration (MCHC), platelet count (PC), total leucocytes (WBC), differential leukocyte count (DLC), reticulocyte count (RC) and blood coagulation using coagulation analyser (activated partial thromboplastin time—APTT, Prothrombin time—PT) and

(continued)

Observation		Monitoring period	Specification
			clotting time by the capillary method
			Electrolytes: Total calcium, potassium, sodium, chloride, phosphorus, inorganic calcium
			Clinical biochemistry: Carbohydrate metabolism: Glucose Lipid metabolism: Total cholesterol, triglycerides Protein metabolism: Albumin, Total protein **Liver function** Hepatocellular: Serum glutamyl pyruvate aminotransferase (SGPT). Hepatobiliary: γ-glutamyl transferase (SGOT), total bilirubin, total bile acid Renal function: Creatinine and urea
Urine analysis (optional)	:	Last week of the dosing/recovery period	Urine volume, appearance (colour and turbidity), specific gravity, pH, ketone bodies, glucose, nitrite, urobilinogen and blood cells
Broncho-alveolar lavage fluid (BALF) analyses	:	Fluid collected from right lung on 29th day for main group animals and 43rd day for satellite group animals	Assessed for lavaged volume, colour, appearance The lavage fluid is centrifuged (2000 rpm for 10 min at 4 °C) and the supernatant is used for LDH, TP and for alveolar macrophages, lymphocytes, neutrophils and eosinophils estimation
Pathology Necropsy is performed on all survived test systems at completion of the exposure or recovery period and those that were found dead during the experiment			
Gross pathology	:	Detailed examination on 29th day in main groups and on 43rd day in satellite groups	Organs such as brain, spinal cord, pituitary, adrenals, eyes, olfactory bulb, larynx, trachea, thyroid and thymus glands, lungs, heart, aorta, liver, pancreas, oesophagus, stomach, small and large intestines, spleen, bone marrow, kidneys, urinary bladder, testes, epididymis, seminal vesicle, prostate, ovaries, uterus, sciatic nerve

(continued)

Observation		Monitoring period	Specification
Organ weight	:	Immediately after necropsy	Absolute and relative organ weights of adrenals, brain, heart, liver, kidneys, lungs (left lobe including main bronchi), spleen, thymus, gonads, epididymis, uterus are measured
Histopathology	:	Organs mentioned in gross pathology are subjected to histopathological examination for range-finding and main study	Initially, the control and high dose groups are examined. If the organ(s) from the high dose show histopathological changes, organs from the other descending doses are then examined

3.9.5 Data Interpretation

- The mean differences between the control and test item exposed groups (low, mid and high concentrations) should be analysed by biostatistical methods.
- The changes observed in one parameter are correlated with the associated parameters to rule out the treatment-related toxicity (e.g. decrease in feed intake causes a decrease in bodyweight, alters plasma glucose, cholesterol, triglycerides and total proteins, causes ulceration in the stomach and histopathological alterations)
- Based on mortality, morbidity and biologically relevant statistical significance between the control and test item exposed groups, the NOAEC and LOAEC are established (GHS 2013).

3.10 Neurotoxicity

Drugs meant for long-term administrations, should be studied for their effects on neuronal structure and chemical transmission. It also encompasses the investigations on the neurobehavioral functions.

3.10.1 Purpose

- To provide information on the permanent or reversible neuropathological lesions that are likely to arise due to the test item administration
- To establish the NOAEL and LOAEL doses using the dose-and time-response relationships.

3.10.2 Definition

"Neurotoxicity is an adverse change in the structure or function of the nervous system that results from exposure to a chemical, biological or physical agent"—OECD 424 (Test No. 424 n.d., p. 424)

3.10.3 Principle

The test item is administered orally at different dose levels to several groups of healthy animals. Dosing is done for 28/90 days and/1 year or more (chronic). The test systems are monitored for behavioural and neurological abnormalities at prefixed time points. At the end of the dosing period, the test systems are euthanized and the sections of the spinal cord, brain and peripheral nerves are prepared to assess the neurotoxicity potential of the test item.

Procedure (OECD GLP n.d.)		
Species	:	Rat is the preferred test system
Strain	:	Common strains like Sprague Dawley, Wistar Albino
Sex	:	Males and females (Female should be non-pregnant and nulliparous)
Age at dosing	:	<9 weeks old
Acclimatization	:	Minimum 5 days prior to the start of the study
Randomization		Healthy test systems are selected and randomized based on Stratified bodyweight or by random distribution. In any case, the mean differences in bodyweight between the groups should not be statistically significant

Study state experiment:
Groups and treatment

1. Test items should be administered daily till the final day of the study
2. The high dose should be fixed in such a way to impart neurotoxic effects but not death in the test system
3. The groups and treatment template are given below in Table 3.11.

Table 3.11 Experiment design and treatment

Groups	Treatment	Separate study		Combined study with the 28-day study		Combined study with the 90-day study		Combined study with the chronic toxicity study	
		Male	Female	Male	Female	Male	Female	Male	Female
Control	Vehicle	5	5	10	10	15	15	25	25
Low dose	Test item	5	5	10	10	15	15	25	25
Middle dose		5	5	10	10	15	15	25	25
High dose		5	5	10	10	15	15	25	25
Satellite–control	Vehicle	5	5	10	10	15	15	25	25
Satellite–High dose	Test item	5	5	10	10	15	15	25	25

3.10.4 Observations: (Research 2020)

Observation		Monitoring period	Specification
Mortality	:	Twice a day (daily)	Test systems are observed for mortality from the start of acclimatization period and throughout the experiment.
Body weight	:	Once in a week	Bodyweight is recorded from the start of acclimatization day till necropsy
Clinical observation	:	Daily (1–2 h after dosing)	Routine clinical signs
		Weekly once	Complete functional observation battery
Functional test	:	Prior to dosing and at end of dosing/recovery period/ prefixed time points	Functional tests include sensory reactivity to stimuli (e.g. visual, auditory and proprioceptive stimuli, limb grip strength and motor activity), learning and memory and specialized behavioral tests like anxiety, depression etc may be included as needed
Feed Intake	:	Once week	Recorded from the start of acclimatization day till necropsy; except on the day of fasting
Water intake	:	Once week	When the test item is administered through drinking water
Blood parameters	:	At the end of the treatment period/recovery period	**Haematology**: Haemoglobin (HGB), Haematocrit (HCT), mean corpuscular haemoglobin (MCH), total erythrocyte count (RBC), mean corpuscular volume (MCV), platelet count, mean corpuscular haemoglobin concentration (MCHC), total leucocytes (WBC), blood clotting time, reticulocytes, neutrophils, lymphocytes, eosinophils basophils and monocytes
			Electrolytes: Total calcium, potassium, sodium, chloride, inorganic calcium
			Clinical biochemistry: Carbohydrate metabolism: Glucose Lipid metabolism: Total cholesterol, triglycerides Protein metabolism: Total protein, albumin **Liver function** Hepatocellular: Serum glutamyl pyruvate aminotransferase (SGPT)

(continued)

Observation		Monitoring period	Specification
			Hepatobiliary: γ-glutamyl transferase (SGOT), total bilirubin, total bile acid Renal function: creatinine and urea
Urine analysis (optional)	:	Last week of the dosing/ recovery	Urine volume, appearance, specific gravity, pH, ketone bodies, glucose and blood cells
Ophthalmological examination	:	Last week of the dosing/ recovery	Based on specific target organ toxicity
Pathology Necropsy is performed on all survived test systems at end of the dosing or recovery period, and those that express undue stress are found dead during the experiment (OECD 2000)			
Gross pathology	:	Organs examined for all types of neurotoxicity study	The areas of the brain examined are the forebrain, the centre of the cerebrum, the midbrain, the hippocampus, the cerebellum, the pons, the medulla oblongata, the eye with optic nerve and retina, the spinal cord at the cervical and lumbar swellings, the dorsal root ganglia, the dorsal and ventral root fibres, the proximal tibial nerve (at the knee), the proximal sciatic nerve and the calf muscle branches of the tibial nerve
Histopathology	:	Organs mentioned in gross pathology are subjected to histopathological examination	Sections include forebrain, the centre of the cerebrum, sections through hippocampus, midbrain, cerebellum, pons, the eye with the optic nerve, the medulla oblongata, spinal cord at the cervical and lumbar swellings, the dorsal root ganglia, the dorsal and ventral root fibres, the proximal sciatic nerve, proximal tibial nerve (at the knee) and the calf muscle branches of the tibial nerve. The spinal cord and peripheral nerve sections should include both cross or transverse and longitudinal sections Initially, the control and high-dose groups are examined. If the organ (s) from the high dose show histopathological changes, the organs from the other descending doses are then examined

3.10.5 Data Interpretation

- The data obtained should be evaluated in terms of incidence, severity and correlation of neurobehavioral and neuropathological effects.
- NOAEL and LOAEL are to be established based on the above-mentioned indices (GHS 2013).

3.11 Carcinogenicity

Drugs developed for long-term administration in humans like anti-diabetics, anti-hypertensives and so on are generally tested for their carcinogenicity potential. The continuous presence of the drug molecules in the systemic fluids and organs might impart undesirable toxic effects. Carcinogenicity study gives information about the tumorigenic potential of such candidates in rodent models. It also helps in:

- Identifying the target organ(s)' carcinogenicity
- Identifying the time of appearance of a neoplasm
- Characterizing the tumour dose–response relationship
- Establishing no-observed-adverse-effect level (NOAEL) and Benchmark Dose (BMD)
- Extrapolating the data to the human doses (Carcinogenicity (two rodent species) n.d.)

3.11.1 Definition

Carcinogens are chemical substances that induce cancer directly or increase the incidences of cancer upon administration by causing damage to genome or cellular metabolic processes and the testing process is called as "Carcinogenicity study" (Carcinogenicity | AltTox.org n.d.)

3.11.2 Principle

The test is conducted by the administration of test items at different dose levels in several groups of the test systems. The commonly used route of administration is oral; but inhalational and dermal routes are also appropriate, depending on the human intended exposure and nature of the test item. The test systems are examined closely for any toxic effects and also for neoplastic developments. Finally, the animals are necropsied for further investigations. The study duration is usually 24 months (Test No. 451 n.d., p. 451).

Table 3.12 Experiment design and treatment details

Groups	Treatment	Male	Female
Control	Vehicle	50	50
Low dose	Test item	50	50
Middle dose		50	50
High dose		50	50
Satellite–Control	Vehicle	10	10
Satellite–High dose	Test item	10	10

Procedure (OECD GLP n.d.)		
Species	:	Rat is the preferred test system
Strain	:	Common strains like Sprague Dawley, Wistar Albino
Sex	:	Male and female (Female should be nulliparous and non-pregnant)
Age at dosing	:	≤8 weeks old
Animal husbandry	:	Animals should be housed at a temperature of 22 ± 3 °C, relative humidity between 30–70% and 12/12 h light–dark cycle. Animals are housed in grouped cages. Pregnant and lactating females should be housed individually in cages
Acclimatization	:	Minimum 7 days before the start of the experiment
Randomization	:	Healthy test systems are selected and randomized based on random distribution. In any case, the mean differences in bodyweight between the groups should not be statistically significant

Study state experiment
Groups and Treatment:

- Both sexes are used (25 male + 25 female). Females should be nulliparous and non-pregnant
- More number of animals (10 in each sex) is included if interim biological sampling is planned to evaluate the development of neoplasms.
- The highest dose should be fixed to impart toxicity but not mortality.
- The dose level spacing is to be done within two- to fourfold intervals for a best investigation (Table 3.12)

3.11.3 Observation

All the animals are checked for mortality and morbidity periodically and special attention should be paid to ascertain tumour growth (Research 2020).

Observation		Monitoring period	Specification
Mortality	:	Twice in a day (daily)	Test systems are observed for mortality from the start of acclimatization period and throughout the experiment.
Body weight	:	Once in a week	Body weight is recorded from the start of acclimatization day till necropsy
Clinical observation	:	Daily (1–2 h after dosing)	Routine clinical signs
		Once in a week	Complete functional observation battery
Feed intake	:	Daily	Recorded from the start of acclimatization day till necropsy; except on the day of fasting
Water intake	:	Daily	When the test item is administered through drinking water; optional
Blood Parameters	:	At the end of the treatment period/recovery period	**Haematology**: Haemoglobin (HGB), Haematocrit (HCT), total erythrocyte count (RBC), mean corpuscular volume (MCV), mean corpuscular haemoglobin (MCH), mean corpuscular haemoglobin concentration (MCHC), platelet count, total leucocytes (WBC), blood clotting time, reticulocytes, neutrophils, lymphocytes, eosinophils basophils and monocytes
			Electrolytes: Total calcium, potassium, sodium, chloride, inorganic calcium
			Clinical biochemistry: Carbohydrate metabolism: Glucose Lipid metabolism: Total cholesterol, triglycerides Protein metabolism: Total protein, albumin **Liver function** Hepatocellular: serum glutamyl pyruvate aminotransferase (SGPT) Hepatobiliary: γ-glutamyl transferase (SGOT), total bilirubin, total bile acid Renal function: creatinine and urea

(continued)

Observation		Monitoring period	Specification
Urine analysis	:	Last week of the dosing/ recovery	Urine volume, appearance, specific gravity, pH, ketone bodies, glucose and blood cells
Ophthalmological examination	:	Last week of the dosing/ recovery	Based on specific target organ toxicity
Pathology Necropsy is performed on all survived test systems at end of the dosing or recovery period, and those that express undue stress are found dead during the experiment (OECD 2000)			
Gross pathology	:	Organs examined	Organs such as heart, pancreas, stomach, adrenal gland, parathyroid gland, eye, teeth, tongue, aorta, ileum, jejunum, duodenum, peripheral nerve, brain, thymus, pituitary, caecum, lacrimal gland, prostate, thyroid, liver, lung, coagulating gland, salivary glands, colon, trachea, lymph nodes, testis, seminal vesicle urinary bladder mammary gland, ovary, cervix, vagina, skeletal muscle, uterus epididymis, skin, kidney, ureter, urethra, oesophagus, spinal cord, gall bladder, spleen, the section of bone marrow and/or a fresh bone marrow aspirate, Harderian gland, rectum
Organ weight	:	Immediately after necropsy	Absolute and relative organ weight of brain, heart, liver, paired kidneys, paired adrenals, spleen, paired testes, paired epididymis, male sex glands, uterus with the cervix, paired ovaries, thymus and spleen are measured
Histopathology	:	Organs mentioned in gross pathology are subjected to histopathological examination	Initially, control and high-dose groups are examined. If the organ (s) from the high dose show histopathological changes, organs from the other descending doses are then examined

3.11.4 Data Interpretation

- Dose-dependent incidences of tumorigenicity or frequency of cancer should be assessed in test item administered test systems.
- NOAEL and LOAEL should also have arrived

Table 3.13 GHS classification criteria for carcinogenicity

Category	Criteria
Category 1A	Known to have carcinogenic potential for humans.
Category 1B	Presumed human carcinogens
Category 2	Suspected human carcinogens

- Based on NOAEL, LOAEL and BMD the test item is classified (GHS 2013) (Table 3.13).

3.12 Reproduction and Developmental Toxicity

3.12.1 Purpose

Reproductive and developmental toxicology studies provide information on risk and health hazards likely to arise following exposure to the test item during the crucial period of sexual maturity/mating, pregnancy, lactation and offspring growth using male and female rats.

3.12.2 Definition

- "Reproduction toxicity represents harmful effects on the progeny and/or an impairment of male and female reproductive functions or capacity"—OECD
- "Developmental toxicity: the manifestation of reproductive toxicity, representing pre-, peri- post-natal, structural or functional disorders in the progeny"—OECD (Test No. 421 n.d., p. 421)

Reproductive and developmental toxicity screening involves a series of procedures carried out for a test item, in three segments to assess the changes in the animals' gonadal structure and functions, fertility, conception, development of the conceptus, parturition and lactation.

On the basis of the reproductive and development cycle of the test system, the study is carried out in the following segments:

- Segment I—Fertility and reproductive performance study
- Segment II—Prenatal developmental toxicity study
- Segment III—Extended One-Generation Reproductive Toxicity Study (EOGRTS)

3.13 Fertility and Reproductive Performance Study (Segment I)

3.13.1 Purpose

Segment I of the reproductive toxicology study provides information on the changes occurring in the reproductive performances and the fertility patterns following test item administration at different dose levels in several groups of male and female rats

3.13.2 Principle

Male and female rats are used for the segment-I study.

In males, the test item is given 2 weeks before mating, during the mating period and 2 weeks post-mating.

In females, the test item is administered 2 weeks prior to mating (covering a minimum of two oestrous cycles), 2 weeks during mating days, 21/22 days gestation and 13 days lactation.

- During test item administration period, the animals are observed for the development of clinical signs, if any. At the end of the dosing period, all the survived animals are euthanized and necropsied to determine the male (e.g. on mating behaviour, libido, epididymal sperm maturation, capacitation) and female (oestrous cycle [ovarian and uterine effects], development of pre-implantation stages of the implantation and embryo, tubal transport,) specific endpoints, clinical chemistry and histopathological changes (Hood and Parker 2008)

Procedure (OECD GLP n.d.)		
Species	:	Rat
Strain	:	Sprague Dawley, Wistar Albino
Sex	:	Male and female (nulliparous, nonpregnant with normal oestrous cycle)
Age at dosing	:	8–10 weeks
Animal husbandry	:	Animals should be housed at a temperature of 22 ± 3 °C, relative humidity between 30–70% and 12/12 h light–dark cycle. Animals are housed in grouped cages. Pregnant and lactating females should be housed individually in cages
Acclimatization	:	Minimum 5 days before start of the experiment
Randomization	:	Healthy males and females tests are selected and randomized based on stratified bodyweight or by random distribution. In any case, the mean differences in body weights between the groups should not be statistically significant.
No of animals per dose level	:	Male: 10/dose.

Study state experiment (Test No. 421 n.d., p. 421)
Dosing regimen

1. The study includes control and three test groups.
2. A minimum of three dose levels of the test item should be selected for the study. The highest dose is selected on the basis of acute, sub-acute and chronic toxicity studies. The highest dose should not produce lethality but is expected to impart mild to moderate fecundity or developmental defects in rats. The test item is administered daily, for all 7 days, via the same route as intended for human exposure.
3. In the case of the male, the test item is administered 2 weeks before mating, during mating and 2 weeks after mating, followed by scheduled necropsy.
4. Females selected for the study should be assessed for at least four to five oestrous cycles before exposure to the test item. Females should be dosed 2 weeks prior to mating (ensure to cover minimum two complete estrous cycles) and continues up to day 21/22 of the gestation period, and at least 2 weeks after delivery. Females showing no evidence of copulation should be euthanized between 24 and 26 days after introducing to mating, up to and as well as the day before necropsy.
5. Mating procedure: In the post-treatment, the female is introduced into male cage (overnight) in the ratio of 1:1. The pairing should be maintained the same throughout the study unless any death is recorded. Pregnancy is confirmed by examining (morning time) the sperm-positive vaginal lavage or copulatory plugs in the vagina or cage pan. Confirmation of sperm or copulatory plug is declared as gestation day 0 and the pregnants are housed in individual cages.
6. Litter size fixing: On postnatal day (PND) 4, 4–5 pups/sex should be selected and kept along with the dam exposed for dosing. The blood sample from the surplus pups should be collected for T4 level estimation.

3.13.3 Observations

1. During the test item administration, the animals should be observed once daily for mortality and routine clinical signs (Research 2020)
2. Males and females should be monitored for bodyweight changes, feed intake and functional battery observation on weekly basis. Pregnant rats should be weighted on days 0, 7, 14, 20, 24 h postpartum and 4 and 13 PND
3. Post-treatment, both male and female subjects are subjected to necropsy oecd_gd19.pdf (2000).

Male-specific endpoints

1. Prior to necropsy, the mating behaviour including the animal's libido, mounting behaviour, erection, ejaculation and changes in production and secretion of hormones should be recorded.
2. Sperm parameters including count, quality, viability, morphology, mortality, transport and maturation are recorded.

3. During necropsy, the male sex organs testes, epididymides, seminal vesicles, prostrate and pituitary glands should be subjected to gross and histopathological examination

Female-specific endpoints

1. In case the female test subject(s) is found in moribund condition or dead after mating, they should be necropsied in order to determine the cause. Signs of pregnancy or development of embryo should also be recorded at this stage.
2. Female sex organs uterus, cervix, the vagina should be examined for gross and histopathological changes, if any.
3. The uterine contents should be checked for evidences of implantation. The number of corpus lutea for each ovary should be counted.
4. Dead pups or pups euthanized on PND 13 should also be necropsied and subjected to gross and histopathological examination.

3.13.4 Interpretation

Individual animal data should be critically analysed and the observations on mortality, morbidity, body weight changes, clinical sign abnormalities, gross and histopathological data are compared between control and treated groups. Based on the data interpretations, the NOAEL and LOAEL value of the test item is established (GHS 2013).

3.14 Prenatal Developmental Toxicity Study (Segment II)

3.14.1 Purpose

- The segment II study provides information on the risk and health hazards likely to arise following exposure to the test item in confirmed pregnant rats from gestation day (GD) 5 to 13/20.
- The study provides information on the impact of the test items on pregnant rats health, foetus organogenesis and growth

3.14.2 Definition

Developmental toxicology: "the study of adverse effects on the developing organism that may result from exposure prior to conception, during prenatal development, or postnatally to the time of sexual maturation. The major manifestations of developmental toxicity include (1) death of the organism, (2) structural abnormality, (3) altered growth and (4) functional deficiency. Developmental toxicology was formerly often referred to as teratology"—OECD (Test No. 414 n.d.)

3.14.3 Principle

In the segment-II study, the test item is administered at three dose levels in several groups of pregnant rats via the route intended for human administration. The test item is administered from GD-5 to GD-20. At the end of the treatment duration, the pregnant rats are euthanized and the uterine contents are examined for implantation, resorption (if any), no of alive and dead foetus. Based on the dose-related response relationship, the no-observed-adverse-effect level (NOAEL) or benchmark dose are established (Derelanko and Auletta 2014)

Procedure (OECD GLP n.d.)		
Species	:	Rat
Strain	:	Sprague Dawley, Wistar Albino
Sex	:	Pregnant female
Age at dosing	:	8–10 weeks
Animal Husbandry	:	Animals should be housed at a temperature of 22 ± 3 °C, relative humidity between 30–70% and 12/12 h light–dark cycle. Animals are housed in grouped cages. Pregnant and lactating females should be housed individually in cages
Acclimatization	:	Minimum 5 days before the start of the experiment
Randomization	:	Healthy pregnant rats should be selected and randomized based on bodyweight
No of animals	:	Pregnant female rats: 20/dose

Study state experiment (Test No. 414 n.d.)

1. The confirmed pregnant rats should be placed in individual cages.
2. The test item is administered at three dose levels to several groups of pregnant rats from gestation day (GD) 5 to GD 20 (or scheduled caesarean section) and concurrent control groups treated with vehicle are handed in the same manner as test item treated groups.
3. The highest test dose should be selected in such a way that it produces mild to moderate maternal and/or developmental toxicity. The descending sequel doses are spaced apart (two- to fourfold interval) in such a way that the study supports to establish NOAEL and BMD

3.14.4 Observations

1. Pregnant rats should be examined twice daily for mortality and the clinical signs of toxicity, if any.
2. Body weight should be recorded on GD 0, and every 3 days during the dosing period, and once before the scheduled caesarean section.
3. Feed intake should be measured every third day during the dosing period
4. Female rats showing signs of morbidity or abortion should be sacrificed humanely and subjected to macroscopic evaluation. (OECD 2000)

5. Dams die during the treatment period should be subjected to gross pathological examination. Organs showing gross pathological changes should be subjected to histopathologic examination
6. At the end of the treatment duration, the below-mentioned observations should be recorded in the dam and foetus (Research 2020)

3.14.5 Maternal Endpoints

(a) Gestation length
(b) Before the scheduled caesarean section, the blood samples should be collected and measured for T3, T4 and thyroid-stimulating hormones levels to ascertain the effects of the test items on endocrine functions.
(c) The gravid uterus should be collected and weighed separately and subjected to histopathology.
(d) Detailed examination of the uterus should be carried out to check the number of implant sites and corpora lutea per dam
(e) Number and percentage of living, dead and resorbed foetus.
(f) The number and percentage of affected implants (non-live + malformed), total resorptions and stillbirths per litter
(g) Number of live foetuses
(h) The absolute weight of all components of reproductive organs should be recorded

3.14.5.1 Foetal Endpoints

The uterine contents should be collected and observed for signs of toxicity. Each foetus is separated from the uterus and the umbilical cords are cut off and blotted on absorbent paper in order to remove the traces of blood and amniotic fluid. The following foetal specific endpoints should be recorded

(a) Number of live and dead foetuses
(b) Live foetus body weight
(c) Sex ratio per litter (Fig. 3.18)
(d) The number of foetus with external and visceral malformations per litter

3.14.6 Interpretation

Individual animal data on the mortality, morbidity, number of abortions, live and dead foetus, clinical chemistry of pups and litter, gross and histopathological changes are compared between the vehicle and the test item groups. Based on the dose-related maternal and developmental toxicity the NOAEL and LOAEL of the test item are established (GHS 2013).

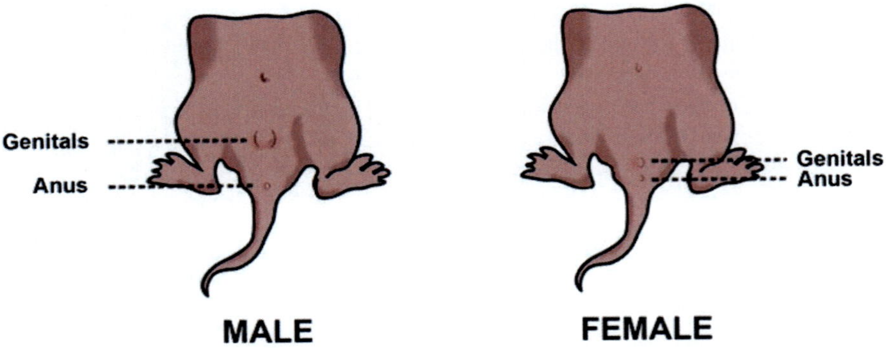

Fig. 3.18 Sex determinations in new-born rodents: The sex of the foetus is determined by measuring the anogenital distance (AGD), that is the distance from the anus to the genitals. The distance is longer at birth in males than females

3.15 Extended One-Generation Reproductive Toxicity Study (EOGRTS) (Segment III)

3.15.1 Purpose

- EOGRTS or segment III study gives information about the health hazards likely to arise in various life stages following the exposure to the test item prior to conceptus [in parent males and females], during in-utero development of embryo/foetus, preweaning (lactation) (in pregnant and dams), adolescence and adulthood (first generation offspring (F1)).
- The effects of the test item in terms of F1 generation reproductive functions, development of nervous and immune systems is investigated to establish the NOAEL, LOAEL and benchmark dose

3.15.2 Principle

The test item is administered at three or four dose levels to several groups of adult males and females through the human intended route of exposure. The test item is administered to parent males and females for a period of 10 weeks (2 weeks prior to mating, 2 weeks during mating, 6 weeks after-mating [covering pregnancy and lactation period in females]). The offspring (F1 generation) produced are exposed to test items from weaning till adulthood. Clinical observations include feed intake, body weight, signs of toxicity, male and female reproductive functions, growth and development of foetus and offspring's sexual maturation, nervous and immune systems development. The detailed gross and histopathological examination of visceral organs is also performed to establish the NOAEL and LOAEL. (Collins 2006)

Experimental condition for exposure of test item in Parental rats

Procedure (OECD GLP n.d.)		
Species	:	Rat
Strain	:	Sprague Dawley, Wistar Albino
Sex	:	Male and female
Age at dosing	:	Within sex, same age (approx. 90 days old) and sexually matured. Females should be nulliparous and non-pregnant. Further, females should be checked for the regular oestrous cycle (4–5) for at least three to four cycles prior to the allocation to treatment
Animal husbandry	:	Animals should be maintained at a temperature of 22 ± 3 °C, relative humidity between 30–70% and 12/12 h light–dark cycle
Acclimatization	:	Minimum 5 days before the start of the experiment
Randomization	:	Healthy pregnant rats should be selected and randomized based on body weight
Housing during dosing	:	Animals are housed in grouped cages during the dosing period
Housing during the mating period	:	Female is introduced into male cage at a 1:1 ratio and housed individually
Post-mating male	:	Housed individually
Pregnant and lactating females	:	Housed individually/mother with litters
Randomization	:	Males and females are randomized based on stratified body weight
No of animals per dose level	:	Min 25 males and 25 Females are selected and co-habituated to get a minimum of 20 pregnant rats/dose level

Study state experiment (Test No. 443 n.d.)

1. Sexually matured parent males and females are administered with the test item. Three or four dose levels are selected and administered to different groups of animals.
2. Dose levels should be selected based on the repeated oral toxicity, toxicokinetics and so on. The highest dose should be selected with an aim to potentially affect fertility, pregnancy in P generation and growth and development of F1 offspring. The descending doses sequel should be spaced at two- or fourfold intervals
3. Parent males and females are dosed with the test item for a period of 2 weeks before to mating
4. Then, the pretreated females are introduced into the males (pretreated) cage at a 1:1 ratio and the dosing is continued for 2 weeks in both sexes
5. Pregnancy is confirmed by examining (morning time) the sperm-positive vaginal lavage or copulatory plugs in the vagina or in the cage pan. Confirmation of sperm or copulatory plug is declared as gestation day 0 and the pregnant rats are housed in individual cages. Dosing is then continued till the end of lactation period in females
6. Following confirmation of pregnancy, the male should be separated and housed individually and dosing continued for the next 6 weeks (until F1 weaning). Similarly, the female should be dosed till the completion of weaning.

7. On post-natal day 4, the pups are adjusted to select five males and five females and housed with the litter and the blood sample is collected from the surplus pups, pooled and used for biochemical estimation including T4 and TSH hormones
8. During the entire experimental period viz. pre-mating, mating, gestation, lactation and post-mating, the parent animals should be observed for mortality and moribundity (twice daily), body weight (weekly), feed intake (alternate day) and complete functional observation battery once in a week
9. Date of introduction to mating, date of pregnancy confirmation and gestation length should also be recorded. Signs of difficulty and prolonged parturition should also be recorded in females
10. At the end of the pre-determined dosing period, blood samples were collected from the parent males and females and subjected to gross and histopathological examination (detail of terminal observations are given in Table 3.15) oecd_gd19.pdf (2000).

Experimental condition for exposure of test item in F1 Offspring

Procedure		
Age	:	Immediately after weaning (around post-natal day 21)
Husbandry	:	Animals should be maintained at a temperature of 22 ± 3 °C, relative humidity between 30–70% and 12/12 h light–dark cycle
Dosing period	:	From weaning till adulthood
Randomization	:	On post-natal day 21, 20/sex/dose group (one male and one female/litter) are randomly selected and assigned to one of the below-mentioned cohort study Table 3.14

3.15.3 Steady State Experiment

Table 3.14 Cohort studies in F1 offspring generation

Study	Extension	Investigation	No of animals	Sexual maturation	Dosing period
Cohort 1	A	Assessment of effects of test item on reproductive organs development and general toxicity	40/dose (20/sex [1 male + 1 female/ litter]/dose)	Yes	Immediate after weaning till PND 75–85; once daily
	B	Assessment of reproductive performance by mating F1 animals	40/dose (20/sex [1 male + 1 female/ litter]/dose)	Yes	Immediate after weaning till PND 140; once daily. PND -175, if triggered for F2 generation; once daily

(continued)

Table 3.14 (continued)

Study	Extension	Investigation	No of animals	Sexual maturation	Dosing period
Cohort 2	A	Assessment of neurobehavioral functions and neurohistopathological examination	20/dose (20/sex [1 male + 1 female/ litter]/dose)	Yes	Immediate after weaning till PND 76–89, once daily
	B	Examination of neurohistopathological on PND 21	20/dose (10 males + 10 females [1 male + 1 female/ litter])	No	Immediate after weaning till PND 21/22
Cohort 3	–	Assessment of immune system function including in the T-cell dependant antibody response determination (TDAR) at PND 56 ± 3.	20/dose (20/sex [1 male +1 female/ litter]/dose)	Yes	Immediate after weaning till PND 56

3.15.3.1 Cohort—1A: Reproductive toxicity Assessment

(a) Immediately following delivery, the number and sex of pups, livebirth, stillbirth, gross anomalies like cleft palate, abnormal skin texture and colour and so on should be recorded in the pups
(b) The pups should be weighed on PND—0, 4,7,14 and 21
(c) Clinical signs such as skin texture, eyes, the occurrence of secretions, posture and locomotor functions and so on should be recorded during weighing or on a case-to-case basis
(d) Anogenital distance (AGD) should be determined from PND 0 to PND 4. The balano-preputial separation day in males and vaginal patency day in females should be compared between control and test item administered rats.
(e) The terminal observations are given in Table 3.15

3.15.3.2 Cohort—1B: Reproductive performance Assessment

Fl males and females of the same dose group are cohabited (siblings pairing is avoided) for 2 weeks. The treatment with the test item is continued till PND 90–120. The generation of F2 offspring is similar as per the procedure followed in the parent generation.

3.15.3.3 Cohort—2: Neurotoxicity Assessment

Cohort—2A: At PND 76–89, the F1 offspring should be evaluated for sensory functions such as startle response (PND 24 ± 1), functional behavioural

Table 3.15 Terminal Observations in parental and F1 offspring

Observation		Generation	
		Parent	F1
Clinical chemistry	:	• Parent: At the end of F1 weaning • F1 offspring: At the end of the assigned cohort study • In case of surplus pups, the blood sampling is performed on PND 4 and pooled for clinical chemistry and then subjected to gross and histopathology • In case of cohort 1A, the blood sampling is performed on PND 21/22 and subjected to gross and histopathology (those not selected for cohort) Carbohydrate metabolism: Glucose Lipid metabolism: Total cholesterol, triglycerides Protein metabolism: Total protein, albumin **Liver function** Hepatocellular: serum glutamyl pyruvate aminotransferase (SGPT) Hepatobiliary: γ-glutamyl transferase (SGOT), total bilirubin, total bile acids Renal function: creatinine and urea Endocrine disruptor markers: T4 and TSH	
Haematology	:	• Parent: At the end of F1 weaning • F1 offspring: At the end of the assigned cohort study Haematocrit (HCT), haemoglobin (HGB), total erythrocyte count (RBC), mean corpuscular haemoglobin (MCH), mean corpuscular volume (MCV), mean corpuscular haemoglobin concentration (MCHC), platelet count, total leucocytes (WBC), blood clotting time, reticulocytes, neutrophils, lymphocytes, eosinophils, basophils and monocytes	
Urine analysis	:	Urine samples from the adults should be obtained at the termination of the study using metabolic cages. In parent and adult F1 generation the urine volume, appearance, specific gravity, pH, ketone bodies, glucose and blood cells should be determined	
Sperm parameters	:	• Parent: At the end of F1 weaning • F1 offspring: At the end of the assigned cohort study • Sperm parameters: Motility, morphology (fusion, isolated heads, misshapen heads/tails) and so on.	
Gross pathology	:	At the end of the study, the animals from the P and cohort 1A and 1B in F1 should be examined for gross pathological changes with special attention on reproductive organs Brain, pituitary, the eye with the optic nerve, lungs, thymus, thyroid, heart, liver, spleen (including splenic lymphocyte subpopulation [B lymphocytes, CD4+ and CD8+ T lymphocytes and natural killer cells] analysis), lymph nodes (one proximal and one distal from the route of test item administration), kidneys, adrenal glands, peripheral nerves, the muscle should be harvested. Reproductive organs testes, epididymides and seminal vesicles with coagulation gland, and prostate should be collected and weighed. Oestrous cycles should also be examined on the day of necropsy	
Histopathology	:	The organs from the control and high dose groups should be first examined for histopathological changes. If toxic changes are observed in high dose group, lower doses should also be examined to establish the LOAEL and NOAEL	

responses (PND 63–75), locomotor functions and finally subjected to neuropathological assessments

Cohort—2B: Immediate after weaning (PND 21/22), the offspring are euthanized and subjected to complete neuropathological examination

3.15.3.4 Cohort—3: Immunotoxicity Assessment

At PND 56 ± 3, the offspring should be checked for T-cell dependant anti-body responses. The immunity response is evaluated either by measuring plaque-forming cells in the spleen or by measuring sheep red blood cells or Keyhole Limpet Hemocyanin triggered IgM production using the ELISA technique. A decrease in IgM production in the offspring indicates the immunotoxicity of the test item.

3.15.4 Data Interpretation

1. Observations on the body weight, feed intake, haematology, blood biochemistry, clinical signs, gross and histopathology should be compared between the control and test groups
2. Intense focus on the changes that occur during the reproductive phase of either male and female rats, presence or absence of abnormalities in foetus owing to the teratogenicity of the test item and its effects on developmental and growth and finally F1 offspring reproductive should be analysed.
3. Based on the above-mentioned data observations and analyses, the LOAEL and NOAEL items should be established for the test item. (GHS 2013)

3.16 Bacterial Reverse Mutation Test

3.16.1 Purpose

To assess the test item potential to induce point mutations viz. addition, deletion or substitution of one or more DNA base pairs in bacterial strains

3.16.2 Definition

Bacterial reverse mutation test (also called as AMES test) is a non-mammalian genetic toxicology assay conducted to determine the potential of the test item to provoke point mutations in amino acid requiring strains of *Salmonella typhimurium* and/or *Escherichia coli* (Nutrition 2019)

3.16.3 Principle

Bacterial reverse mutation test helps to detect the mutations which revert the mutation present in the tester strain, thus restoring its functional ability to produce an essential amino acid. The revertants grow even in the lack of amino acid required by the parent test strain. The test employs the use of exogenous metabolic activation systems to mimic the mammalian *in vivo* conditions. Bacterial suspensions are exposed to test items in the presence and absence of metabolic activation systems by the Plate incorporation method or pre-incubation method. Other less commonly used methods are the fluctuation and suspension methods (Hamel et al. 2016).

Procedure (Test No. 471 n.d.)		
Bacterial strain	:	*S. typhimurium* TA1535, and
		S. typhimurium TA1537 or TA97 or TA97a, and
		S. typhimurium TA98, and
		S. typhimurium TA100, and
		E. coli WP2 uvrA, or *E. coli* WP2 uvrA (pKM101), or
		S. typhimurium TA102.
Growth phase and titre	:	Approximately, 10^9 cells per mL (titre demonstrated by determination of viable cells using plating experiment, along with each testing or a historical control data on growth curve)

Genotypic characterization of tester strains

Strain	Reversion event	Histidine mutation	LPS defect	R-factor	Plasmid
TA98	Frame shift	D3052	rfa, uvrB	+R	pKM101
TA100	Base pair substitution	G46	rfa, uvrB	+R	pKM101
TA1535	Base pair substitution	G46	rfa, uvrB	−R	–
TA1537	Frame shift	C3076	rfa, uvrB	–	–
TA102	Transition/ transversion	G428	rfa, uvrB	+R	pKM101, pAQ1
E.coli (WP2 uvrA)	Base pair substitution	–	uvrA	–	–

Medium	:	Minimal basal agar medium (Vogel–Bonner medium) Overlay agar with histidine and biotin or tryptophan to support few cell divisions of bacteria or growth of spontaneous revertants
Metabolic activation	:	Cofactor supplemented post-mitochondrial fraction (S9) prepared from the rat liver, in which enzyme induction is activated by Aroclor 1254 or a combination of

Table 3.16 Types of strains used as the positive control

Strain	Positive control	
	+S9	−S9
TA98	2-Aminoanthracene	2-Nitrofluorene [CAS no. 607-57-8
TA100 and TA1535		Sodium azide [CAS no. 26628-22-8
TA1537		9-Aminoacridine [CAS no. 90-45-9]
TA102		Cumene hydroperoxide [CAS no. 80-15-9]
E.coli (WP2uvrA)		4-nitroquinoline1-oxide [CAS no. 56-57-5]

β-naphthoflavone and phenobarbital. S9 is used at concentrations from 5 to 30% v/v in the S9-mix.

Controls : Negative control
Positive control (Strain-specific) (Table 3.16)

Test item concentration: Soluble non-cytotoxic test item—5 mg/plate or 5 μL/plate
Non-soluble non-cytotoxic test item—Lowest precipitating dose
Cytotoxic Test Item—Highest dose should show mild to moderate degree of toxicity

3.16.4 Study State Experiment

Experiment

Preliminary toxicity
(Infers cytotoxic profile of Test item)
Employs only TA100 and/or Ta1535

Mutation assay
(Infers mutagenic profile of test item)
Employs all five strains

Plate Incorporation Method

About 0.05–0.1 mL of test item, 0.1 mL of bacterial suspension, 0.5 mL of S9 mix (presence of metabolic activation) or sterile buffer (absence of metabolic activation) should be mixed with 2 mL of overlay agar and plated onto the minimal agar plate.

Pre-incubation Method

1. About 0.05–0.1 mL of test item should be pre-incubated with 0.1 mL of bacterial suspension, 0.5 mL of S9 mix (presence of metabolic activation) or sterile buffer (absence of metabolic activation) for a period of approximately 20 min at 30–37°C and tubes should be aerated by incubating in a shaker. The contents will be mixed with 2 mL of overlay agar and plated onto the minimal agar plate.
2. The test should be performed in triplicates.
3. Plates should be incubated at 37°C for 48–72 h
4. Plates should be counted for revertant colonies and results will be recorded.

3.16.5 Data and Reporting

Data are expressed as revertant colonies per plate. Single plate count and mean number of revertant colonies with standard deviation should be presented.

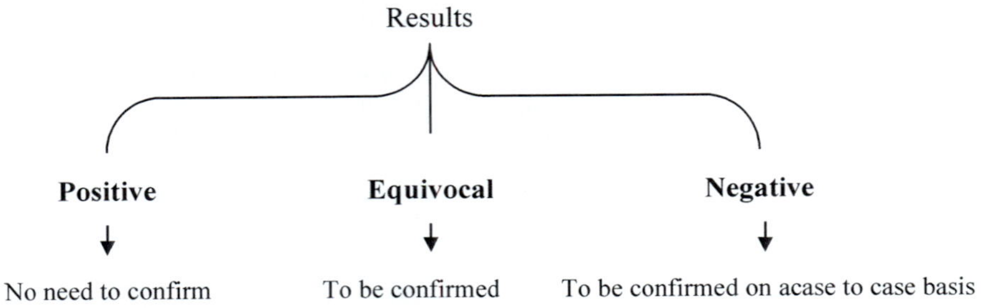

Confirmatory assay: Study parameters are modified for example, changes in concentration levels, treatment approach (plate incorporation or liquid preincubation) and metabolic activation conditions (Research 2020).

3.16.6 Interpretation

1. Negative control data should be within the range of previous control data collected in-house.
2. Positive control data should be within the in-house historical control data, which validates the test.
3. The revertant colonies of the test item should be compared with negative control data statistically.
4. Biological relevance should also be determined.
5. A positive response is defined as a concentration-dependent increase in the revertant colonies when compared to the negative control data, that is two to threefold increases in the revertant colonies.

6. Positive findings in a bacterial reverse mutation test indicate that the test item's ability to induce point mutations in the genomes of *Salmonella typhimurium* and/or *Escherichia coli* by causing base substitutions or frameshifts.
7. Negative findings indicate that the test substance is not mutagenic in the tested strains and conditions (GHS 2013).

3.17 In Vivo Mammalian Erythrocyte Micronucleus Test

In the nineteenth century, Howell–Jolly identified the micronuclei (also called as Howell–Jolly bodies) in cat and rat erythrocytes (Hayashi 2016). In vivo rodent micronucleus assay is one of the batteries of genotoxicity assays, which is widely accepted by various drugs and chemicals regulatory bodies, to assess the carcinogen potential of a test item. Results of micronuclei assay are considered as the best score to estimate the carcinogenicity potential of the test item when compared to other genotoxicity assays.

3.17.1 Purpose

This test is used to determine the potential of a test item to produce micronuclei in rodent's bone marrow erythropoietic cells and to classify as carcinogen or non-carcinogen

3.17.2 Definition

The test used to determine the potential of a test item to produce micronuclei as a result of chromosomal fragments lagging (clastogenicity) or full chromosome lagging (aneugenicity) is known micronuclei assay (Nutrition 2019) (Fig. 3.19).

3.17.3 Principle

The test item is administered at different dose levels to the experimental animals in a single dose or multiple doses through the human intended route and the blood or bone marrow sample is collected at a predetermined time (24 and/48 h). The samples are stained and examined for the presence of micronuclei using a microscope and image analysis. The test is considered positive when the treatment group(s) shows a significant increase in the frequency of micronucleated immature erythrocytes compared to the negative control.

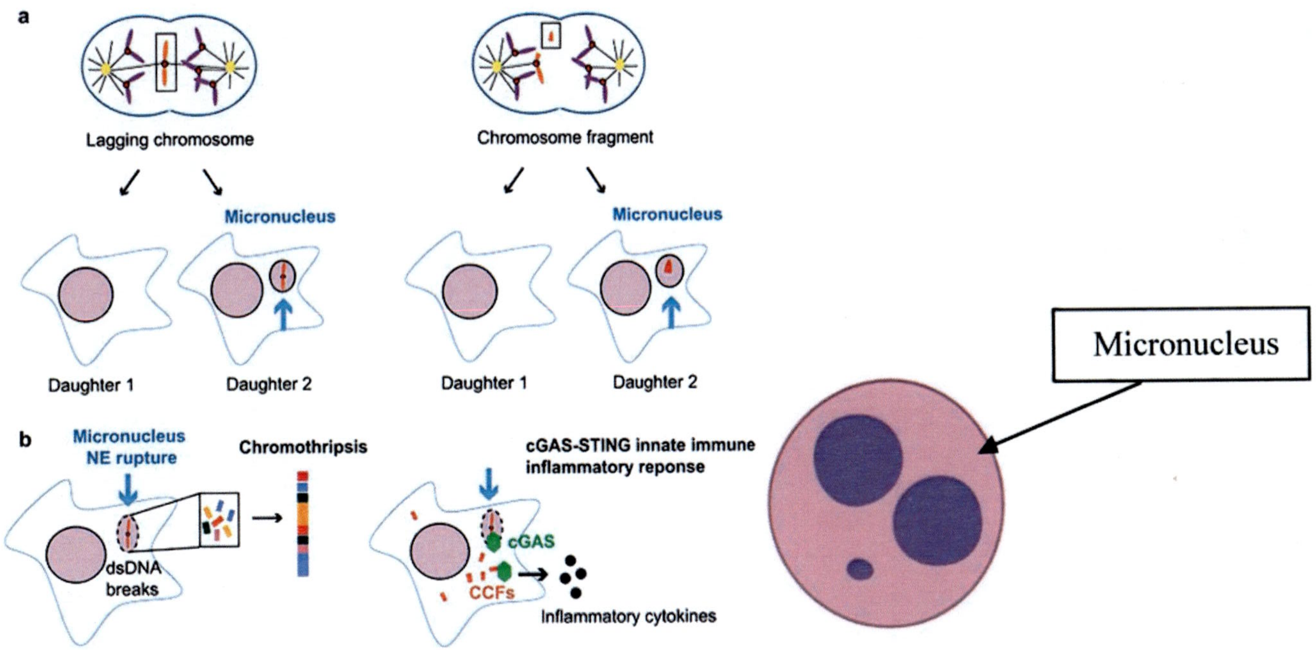

Fig. 3.19 Consequences of Micronucleus formation. [Adapted with modifications from Kwon et al. (2020)]

Procedure (OECD GLP n.d.)		
Preferred species	:	Rat or mouse
Strain	:	Commonly used strains
Sex	:	Male and female (Female should be nulliparous and non-pregnant)
Age at dosing	:	6–10 weeks
Animal Husbandry	:	Animals should be housed at a temperature of 22 ± 3 °C, relative humidity between 30–70% and 12/12 h light–dark cycle. Animals are housed in grouped cages. Pregnant and lactating females should be housed individually in cages
Acclimatization	:	Minimum 5 days before the start of the experiment

Steady-state experiment (Test No. 474 n.d.)

1. The test should be performed with positive and negative controls. Compound (s) used as positive control should produce micronuclei at the exposure levels. The negative control is the vehicle used for test item/positive control preparation and should be used in the test.
2. The test is performed in one of the three ways viz.:
   (a) **Single exposure:** The test item is administered at several dose levels to different groups of animals only once. In the case of bone marrow, first sampling should not be obtained earlier than 24 h and second sampling not beyond 48 h following test item administration. In case of peripheral blood, first sampling should be obtained not earlier than 36 h and second sampling not extending beyond 72 h following test item administration.
   (b) **Two dose exposure and single sampling:** The test item is administered at several dose levels to different groups of animals for two days at 24 intervals. In the case of bone marrow, sampling should be obtained once between 18 and 24 h. In the case of peripheral blood, sampling should be obtained once 36–48 h following the last dose administration.
   (c) **Three dose exposure and single sampling:** The test item is administered at several dose levels to different groups of animals for three days at 24 intervals. In the case of bone marrow, sampling should be obtained not earlier than 24 h. In the case of peripheral blood, sampling should not be collected later than 40 h following the last dose administration.
3. When there is no data available on the test item regarding the genotoxicity potential, a dose range-finding study should be performed using a dose of 2000 mg/kg.
4. The main study is conducted using the minimum of three dose levels. The maximum dose should be selected with an aim to impart toxicity in bone marrow/erythrocytes in animals. The test item is administered through the human intended route of exposure.
5. At the prefixed time point the animals are euthanized, femur bone or blood sampling is collected in a sequential order viz. negative control, positive control, low, mid and high dose groups (OECD 2000).

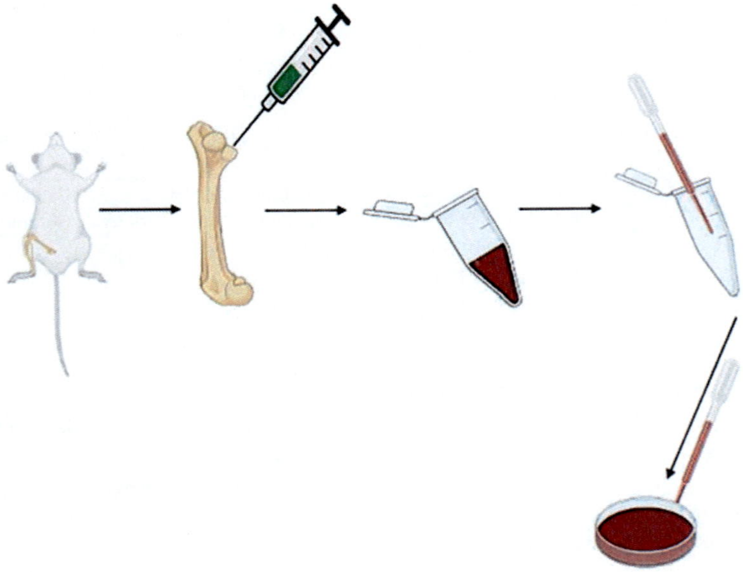

Fig. 3.20 Mouse bone marrow isolation and in vitro culture

6. Following the separation of the femur bone, one end of the bone is scissored to flush out the contents using a syringe loaded with FBS/FCS (Fig. 3.20).
7. The contents are centrifuge at 1600 rpm for 5 min at room temperature and the supernatant is decanted. The cell pellet is re-suspended in the FBS/FCS.
8. The smear of the cells is prepared onto a glass slide and air-dried.
9. Then the slides are fixed using ice-cold methanol for 10 min, softly wash with distilled water and air-dried.
10. The slides are dipped in Giemsa stain (1:5 with water) in a coupling jar and stained for 2–5 min. Then the slides are washed two or three times with distilled water, air-dried and mounted using dibutyl phthalate polystyrene xylene (DPX)
11. Polychromatic erythrocytes (PCE) appear blue to purple in colour and normochromatic erythrocytes (NCE) appear as pink to red colour (Fig. 3.21).
12. In the case of erythrocytes, 2000 total cells to count and in the case of bone marrow 500 cells to be counted. The PCE:NCE ratio should be determined
13. Micronuclei frequency (expressed as percent micronucleated cells) by analysing the number of micronucleated polychromatic erythrocytes (MNPCEs).
14. An increase in the number of micronucleated PCE/increase in micronuclei frequency indicates the mutagenic potential of the test item. On the other hand, a reduction in PCE: NCE ratio indicates cytotoxicity of the test item.

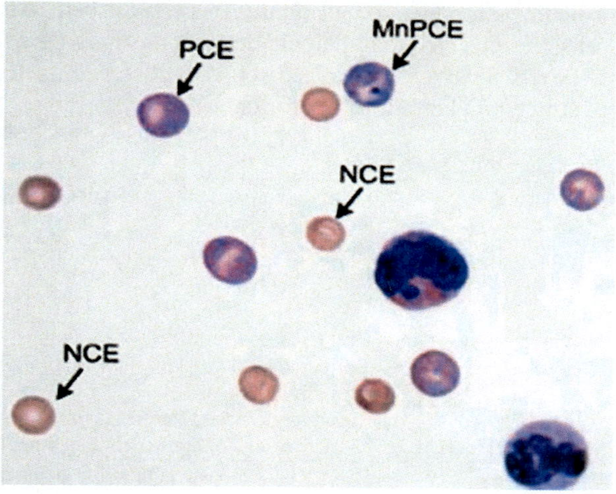

Fig. 3.21 Bone marrow cells: *NCE* normochromatic erythrocytes, *PCE* polychromatic erythrocytes, *MnPCE* micronucleus polychromatic erythrocytes. [Adapted from Salmani et al. (2015)]

3.17.4 Interpretation of Result

A test item is considered to be genotoxic when the treatment group(s) shows a significant increase in the frequency of micronucleated immature erythrocytes compared to the negative control (GHS 2013).

3.18 In Vitro Chromosomal Aberration Test

3.18.1 Purpose

To assess the potential of the test item to induce structural chromosomal aberrations in the cultured mammalian cells

3.18.2 Definition

In vitro chromosomal aberration test is one of the batteries of tests in genetic toxicology study, conducted to determine the potential of the test item to induce structural chromosomal aberrations or mutagenic potential (Clare 2012).

3.18.3 Principle

Cells cultures (human peripheral blood lymphocytes or CHO-K1 cells) are exposed to the test item in the presence and absence of the metabolic activation system, S9. After predetermined intervals of test item exposure, metaphase arresting substance,

Colchicine is added to the cultures to arrest the cells at the metaphase stage. Further, cells are subjected to hypotonic treatment using KCl, harvested and processed using Carnoy's fixative. The stained cells are observed microscopically for the incidence of structural aberrations (Nutrition 2019).

Procedure (OECD GLP n.d.)		
Test system	:	Cultured human peripheral blood lymphocytes Or CHO-K1 cells
Media	:	RPMI-1640 (Human peripheral blood lymphocytes) with FBS F10 (CHO-K1) with FBS
Culture conditions	:	CO_2 level: 5 ± 0.2% Temperature: 37 ± 1°C Humidity: 80–100% (Routinely checked for the stability of a modal chromosome number and absence of mycoplasma contamination for CHO-K1 cells)
Preparation of cultures	:	CHO-K1—Cells are cultured from stock culture, seeded at required density (incubated at conditions specified above) Human peripheral blood lymphocytes: anticoagulant treated whole blood or separated lymphocytes collected from healthy donors are added to culture media with a mitogen (e.g. phytohemagglutinin) (incubated at conditions specified above)
Metabolic activation	:	Cofactor supplemented post-mitochondrial fraction (S9) made from the rat liver, in which enzyme induction is activated by Aroclor 1254 or a combination of β-naphthoflavone and phenobarbital. S9 is used at concentrations from 1 to 10% v/v in the S9-mix
Controls	:	Negative control Positive control (Strain-specific)

Positive control	
+S9	−S9
Benzo(a)pyrene [CAS no. 50-32-8]	Methyl methanesulphonate [CAS no. 66-27-3]
	Ethyl methanesulphonate [CAS no. 62-50-0]
Cyclophosphamide (monohydrate) [CAS no. 50-18-0 (CAS no. 6055-19-2)]	Ethylnitrosourea [CAS no. 759-73-9]
	Mitomycin C [CAS no. 50-07-7]
	4-Nitroquinoline-N-Oxide [CAS no. 56-57-5]

Test item concentration: Soluble non-cytotoxic test item—2 mg/plate or 2 μL/plate (agrochemicals) or 0.5 mg/mL (ICH guidelines for pharmaceuticals).

Non-soluble non-cytotoxic test item—Lowest precipitating dose.

Cytotoxic Test Item—Highest dose should show mild to moderate degree of toxicity at least 45 ± 5% toxicity

3.18.3.1 Study State Experiment (Test No. 473 n.d., p. 473)

→ **Preliminary toxicity:** Short exposure (3-6 h): Presence and absence of metabolic activation)

→ **Mutation assay:** Long exposure (approximately 22 h) [Absence of metabolic activation]

3.18.3.2 Experimental Design

Culture incubated for 24-48h

▼

Treated with test item for 3-6 h (short exposure), Treated continuously till harvest (long exposure)

▼

Cells should be harvested after 1.5 cell cycle length from starting of the treatment

▼

Treated with colchicine

▼

After one to three hours, KClis added to cells for hypotonic treatment

▼

Cells fixed using Carnoy's fixative

▼

Stained and observed microscopically

3.18.3.3 Microscopic Analysis

- Determination of toxicity—Mitotic Index for human peripheral blood lymphocytes
  Reduction in cell counts for CHO-K1 cells

$$\text{Mitotic Index} = [\text{No.of metaphases}/\text{Total no.of cells score}] \times 100$$

- Minimum 200 well spread metaphases are analysed per concentration for the presence of aberrations.
- Cells analysed should have ±2 model chromosome number.

3.18.3.4 Data Analysis

- The experiment unit is a cell.
- The percentage of cells with structural chromosomal aberration(s) should be assessed.
- The number and frequencies of structural chromosomal aberration assay should be evaluated.

- Gaps are recorded and reported independently, but not included in the overall aberration frequency.

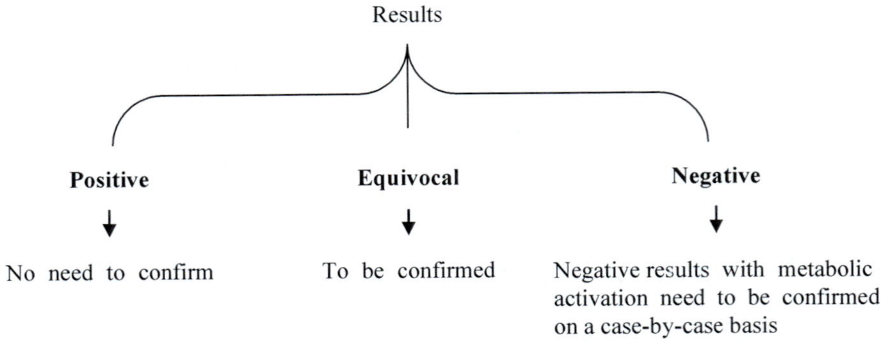

3.18.4 Interpretation

1. Negative control data should be in the range of in-house historical control data.
2. Positive control data should be within in-house historical control data which validates the test.
3. The incidence of structural aberrations of the test item should be compared with the negative control data statistically. (GHS 2013)
4. The biological relevance should also be determined.
5. A positive response is determined by a concentration-dependent increase in the frequency and incidence of cells with chromosomal aberrations, compared to the negative controls.
6. Positive findings indicate the potential of the test item to causes structural chromosomal aberration in cultured somatic cells (Tietje and Brouder 2010, p. 1).

References

Basketter D, Darlenski R, Fluhr JW (2008) Skin irritation and sensitization: mechanisms and new approaches for risk assessment. Skin Pharmacol Physiol 21:191–202

Bei-Bei Z, Jin-Lian H, Yu B, You-Ying Z, Ya-Dong H, Qi X (2016) Safety evaluation of GP-EGF lyophilised powder. J Pharm Biomed Sci 6:295–300

Beshir K, Shibeshi W, Ejigu A, Engidawork E (2017) In vivo Wound Healing Activity of 70% Ethanol Leaf Extract of Becium grandiflorum Lam. (Lamiaceae) in Mice. https://doi.org/10.4314/EPJ.V32I2.3

Carcinogenicity (two rodent species) (n.d.). http://www.fao.org/pesticide-registration-toolkit/registration-tools/data-requirements-and-testing-guidelines/study-detail/en/c/1186760/. Accessed 28 July 2021

Carcinogenicity | AltTox.org (n.d.). http://alttox.org/mapp/toxicity-endpoints-tests/carcinogenicity/. Accessed 28 July 2021

Clare G (2012) The in vitro mammalian chromosome aberration test. In: Parry JM, Parry EM (eds) Genetic toxicology: principles and methods, methods in molecular biology. Springer, New York, NY, pp 69–91

Collins TFX (2006) History and evolution of reproductive and developmental toxicology guidelines. Curr Pharm Des 12:1449–1465. https://doi.org/10.2174/138161206776389813

Derelanko MJ, Auletta CS (eds) (2014) Handbook of toxicology, 3rd edn. CRC Press, Boca Raton. https://doi.org/10.1201/b16632

Design of Dosage Regimens (n.d.). pharmacy180.com. http://www.pharmacy180.com/article/design-of-dosage-regimens-2540/. Accessed 3 Aug 2021

Eaton, D., Gilbert, S., 2013. Principles of toxicology

EPA (n.d.) EPA Health Effects Test Guidelines: Acute Inhalation Toxicity; epa-870-1300.pdf

Hamel A, Roy M, Proudlock R (2016) The bacterial reverse mutation test, pp 79–138. https://doi.org/10.1016/B978-0-12-800764-8.00004-5

GHS (2013) Globally Harmonized System of Classification and Labelling of Chemicals (GHS), Fifth revised edition

GHS (n.d.) GHS Classification Criteria for Skin Corrosion and Irritation

Hayashi M (2016) The micronucleus test—most widely used in vivo genotoxicity test. Genes and Environment 38:18. https://doi.org/10.1186/s41021-016-0044-x

Hayes AW, Wang T, Dixon D (2020) Chapter 13—Toxicologic testing methods. In: Hayes AW, Wang T, Dixon D (eds) Loomis's essentials of toxicology, 5th edn. Academic Press, pp 189–222. https://doi.org/10.1016/B978-0-12-815921-7.00013-2

He J, Tong Y (2019) Effects of silk fibroin moisturizing cream on acetone/ether-induced dry skin pruritus in mice. J Cosmet Dermatol Sci Appl 09:235–243. https://doi.org/10.4236/jcdsa.2019.93021

Hood RD, Parker RM (2008) Reproductive and developmental toxicology. In: Gad SC (ed) Preclinical development handbook. Wiley, Hoboken, NJ, pp 363–422. https://doi.org/10.1002/9780470249055.ch12

Kwon M, Leibowitz ML, Lee J-H (2020) Small but mighty: the causes and consequences of micronucleus rupture. Exp Mol Med 52:1777–1786. https://doi.org/10.1038/s12276-020-00529-z

Lipnick RL, Cotruvo JA, Hill RN, Bruce RD, Stitzel KA, Walker AP, Chu I, Goddard M, Segal L, Springer JA (1995) Comparison of the up-and-down, conventional LD50, and fixed-dose acute toxicity procedures. Food Chem Toxicol 33:223–231. https://doi.org/10.1016/0278-6915(94)00136-c

Nutrition, C. for F.S. and A (2019) Redbook 2000: bacterial reverse mutation test. U.S. Food and Drug Administration. https://www.fda.gov/regulatory-information/search-fda-guidance-documents/redbook-2000-ivc1a-bacterial-reverse-mutation-test. Accessed 28 July 2021

OECD (1987) OECD Guideline for testing of chemicals: acute dermal toxicity. OECD, France. p. 402

OECD (2000) Organization of Economic Cooperation and Development Guidance Document on the Recognition, Assessment, and Use of Clinical Signs As Humane Endpoints for Experimental Animals Used in Safety Evaluation

OECD ENV/JM/MONO (2000)7 (2000)

OECD GLP (n.d.). https://www.oecd.org/chemicalsafety/testing/overview-of-good-laboratory-practice.htm. Accessed 19 July 2021

Research, C. for D.E. (2020) M3(R2) Nonclinical Safety Studies for the Conduct of Human Clinical Trials and Marketing Authorization for Pharmaceuticals. U.S. Food and Drug Administration. https://www.fda.gov/regulatory-information/search-fda-guidance-documents/m3r2-nonclinical-safety-studies-conduct-human-clinical-trials-and-marketing-authorization. Accessed 22 July 2021

Salem H, Katz SA (eds) (2014) Inhalation toxicology, 0th edn. CRC Press. https://doi.org/10.1201/b16781

Salmani A, Kosari A, Pirouzi A, Omidi M, Mohsenzadeh M (2015) Protective effect of methanolic extracts of Thymus vulgaris L. against cyclophosphamide-induced DNA damage in mouse bone marrow cells using the micronucleus test. Trends Pharmacol Sci 1(4):243–250. https://doi.org/10.1111/TIPS.V1I4.47

SCCP (2005) European Commission, Scientific Committee on Consumer products. Memorandum on Classification and categorization of skin sensitisers and grading of test reactions, 2005. sccp_s_01.pdf

Test No. 402: Acute Dermal Toxicity (n.d.). https://www.oecd-ilibrary.org/environment/test-no-402-acute-dermal-toxicity_9789264070585-en. Accessed 29 July 2021

Test No. 403: Acute Inhalation Toxicity | READ online (n.d.).Oecd-ilibrary.org. https://read.oecd-ilibrary.org/environment/test-no-403-acute-inhalation-toxicity_9789264070608-en. Accessed 27 July 2021

Test No. 404: Acute Dermal Irritation/Corrosion (n.d.). https://www.oecd-ilibrary.org/environment/test-no-404-acute-dermal-irritation-corrosion_9789264070622en?__cf_chl_captcha_tk__=pmd_46f9c19dfc7650ec3a62b0a0acb6a3340c5707b0-1627299892-0-gqNtZGzNAw2jcnBszQvi. Accessed 26 July 2021

Test No. 405: Acute Eye Irritation/Corrosion (n.d.). https://www.oecd-ilibrary.org/environment/test-no-405-acute-eye-irritation-corrosion_9789264185333en?__cf_chl_captcha_tk__=pmd_25afc30e81bedc77049f37cf2cb8f35e7e553b66-1627282532-0-gqNtZGzNAw2jcnBszQp6. Accessed 26 July 2021

Test No. 406: Skin Sensitisation (n.d.). https://www.oecd-ilibrary.org/environment/test-no-406-skin-sensitisation_9789264070660-en?__cf_chl_captcha_tk__=pmd_f13d7f2775f6decde9e1740e5d1b8495577e686b-1627362907-0-gqNtZGzNAvijcnBszQv6. Accessed 27 July 2021

Test No. 408: Repeated Dose 90-Day Oral Toxicity Study in Rodents (n.d.). https://www.oecd-ilibrary.org/environment/test-no-408-repeated-dose-90-day-oral-toxicity-study-in-rodents_9789264070707-en. Accessed 27 July 2021

Test No. 412: Subacute Inhalation Toxicity: 28-Day Study (n.d.). https://www.oecd-ilibrary.org/environment/test-no-412-subacute-inhalation-toxicity-28-day-study_9789264070783 en?__cf_chl_captcha_tk__=pmd_f9a903260b5688e1d64509020d0f5bef9518d9bf-1627556092-0-gqNtZGzNAyKjcnBszQv6. Accessed 29 July 2021

Test No. 414: Prenatal Developmental Toxicity Study (n.d.). https://www.oecd-ilibrary.org/environment/test-no-414-prenatal-development-toxicity-study_9789264070820-en. Accessed 30 July 2021

Test No. 421: Reproduction/Developmental Toxicity Screening Test | en | OECD (n.d.). https://www.oecd.org/env/test-no-421-reproduction-developmental-toxicity-screening-test-9789264264380-en.htm. Accessed 29 July 2021

Test No. 423: Acute Oral toxicity - Acute Toxic Class Method | READ online (n.d.). oecd-ilibrary.org. https://read.oecd-ilibrary.org/environment/test-no-423-acute-oral-toxicity-acute-toxic-class-method_9789264071001-en. Accessed 20 July 2021

Test No. 424: Neurotoxicity Study in Rodents | READ online (n.d.). oecd-ilibrary.org. https://read.oecd-ilibrary.org/environment/test-no-424-neurotoxicity-study-in-rodents_9789264071025-en. Accessed 27 July 2021

Test No. 425: Acute Oral Toxicity: Up-and-Down Procedure (n.d.). https://www.oecd-ilibrary.org/environment/test-no-425-acute-oral-toxicity-up-and-down-procedure_9789264071049-en?__cf_chl_captcha_tk__=pmd_ab95c0ea1621308e1d330435afde4b49e1f0eec9-1627110129-0-gqNtZGzNAyKjcnBszQni. Accessed 24 July 2021

Test No. 443: Extended One-Generation Reproductive Toxicity Study | en | OECD (n.d.). https://www.oecd.org/chemicalsafety/test-no-443-extended-one-generation-reproductive-toxicity-study-9789264185371-en.htm. Accessed 30 July 2021

Test No. 451: Carcinogenicity Studies | READ online (n.d.). oecd-ilibrary.org. https://read.oecd-ilibrary.org/environment/test-no-451-carcinogenicity-studies_9789264071186-en. Accessed 28 July 2021

Test No. 471: Bacterial Reverse Mutation Test | READ online (n.d.). oecd-ilibrary.org. https://read.oecd-ilibrary.org/environment/test-no-471-bacterial-reverse-mutation-test_9789264071247-en. Accessed 28 July 2021

Test No. 473: In Vitro Mammalian Chromosomal Aberration Test | en | OECD (n.d.). https://www.oecd.org/env/test-no-473-in-vitro-mammalian-chromosomal-aberration-test-9789264264649-en.htm. Accessed 28 July 2021

Test No. 474: Mammalian Erythrocyte Micronucleus Test (n.d.). https://www.oecd-ilibrary.org/environment/test-no-474-mammalian-erythrocyte-micronucleus-test_9789264224292-en?__cf_chl_captcha_tk__=pmd_977bd6b04fed0659612c0e11aafbe0b4560a2537-1627462654-0-gqNtZGzNAyKjcnBszQvO. Accessed 28 July 2021

Tietje C, Brouder A (eds) (2010) ICH S2R1. In: Handbook of transnational economic governance regimes. Brill | Nijhoff, pp 1041–1053. https://doi.org/10.1163/ej.9789004163300.i-1081.897

Tsai C-H, Wang P-Y, Lin I-C, Huang H, Liu G-S, Tseng C-L (2018) Ocular drug delivery: role of degradable polymeric nanocarriers for ophthalmic application. Int J Mol Sci 19:2830. https://doi.org/10.3390/ijms19092830

Wolff RK (2018) 9.20—Inhalation toxicology studies☆. In: McQueen CA (ed) Comprehensive toxicology, 3rd edn. Elsevier, Oxford, pp 294–314. https://doi.org/10.1016/B978-0-12-801238-3.01955-3

Zaid Alkilani A, McCrudden MTC, Donnelly RF (2015) Transdermal drug delivery: innovative pharmaceutical developments based on disruption of the barrier properties of the stratum corneum. Pharmaceutics 7:438–470. https://doi.org/10.3390/pharmaceutics7040438

Toxicokinetics

4

4.1 Purpose

- Toxicokinetics study provides information (toxicological effects) about the absorption, distribution, metabolism and excretion (ADME) of a test item with respect to time and dose(s) factors. It also provides information on the biodistribution or accumulation of the test item and its metabolites in various biological matrices and the possible mechanisms of toxicity.
- It helps to evaluate the relevance of the systemic toxicity data to predict or extrapolate the health hazards and risks likely to arise in humans with respect to time factor.

4.2 Definition

"Toxicokinetics (TK) is defined as the generation of pharmacokinetic data, either as an integral component in the conduct of non-clinical toxicity studies or in specially designed supportive studies, in order to assess systemic exposure" (S 3 A Toxicokinetics: A Guidance for Assessing Systemic Exposure in Toxicology Studies 2006).

TK is performed using biological matrices like body fluids (plasma/serum), excreta (urine, stool and breath) and peripheral and visceral tissues. It is usually performed as an integral assay in either of the studies like single or repeated dose toxicity study, reproductive toxicity, genetic and carcinogenicity study (Baldrick 2003).

S. B. Chidambaram et al., *Introduction to Toxicological Screening Methods and Good Laboratory Practice*, https://doi.org/10.1007/978-981-16-6092-4_4

4.3 Toxicokinetics in Various Systemic Toxicological Studies

In general, TK study is performed as an integral assay in any one of the below-mentioned systemic toxicity studies and hence as such known as "concomitant Toxicokinetic." In that case, a separate set of animals is allotted and performed.

(a) Single-dose toxicity
(b) Repeated (28/90) dose toxicity
(c) Genetic toxicology
(d) Carcinogenicity study
(e) Reproductive and Developmental toxicity study

Similar to pharmacokinetic study, the biological sampling (blood/tissues) is performed at pre-fixed time points (6–12 time points), following the test item administration and bioanalytical estimation is carried out using HPLC or LC-MS/MS.

If TK is planned in a repeated dose study, the sampling (6–12 time points) will be planned on day 1 or 7 or 14 or 28. Similarly in reproductive (Segment I) and developmental (Segment II and III) toxicity studies, the bio sampling will be planned on the pre-planned segment(s), day and time points (Gad n.d.).

4.4 Toxicokinetics Models

Various Toxicokinetics models include (Fig. 4.1) (Casarett et al. 2008)

1. **Classic toxicokinetics** (traditional model)
   Classic toxicokinetics is the mathematical description, where it relies on the time course of toxicant/test item disposition in the whole organism. In this method, the body is considered as the system of one or more compartments, wherein it does not give organ or tissue data. It is particularly suited for human studies. This model is also known as the "open model".

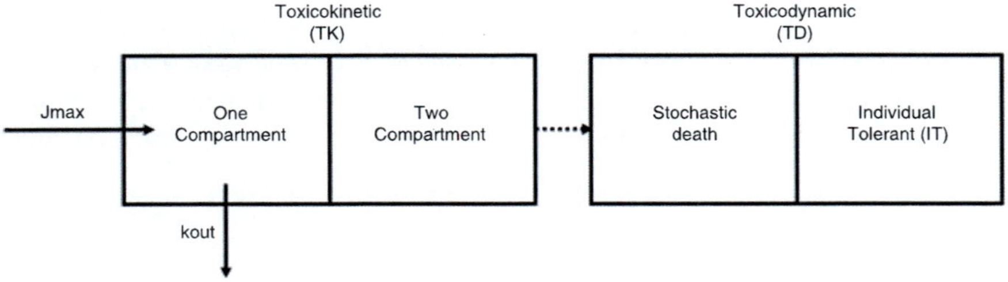

Fig. 4.1 Types of toxicokinetic and toxicodynamic models [adapted with slight modification from Gao et al. 2019]

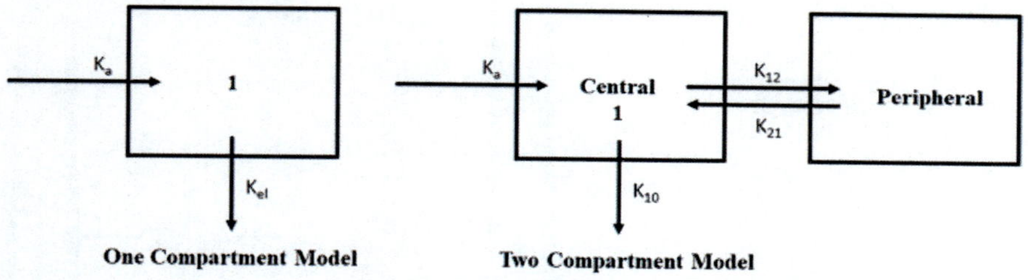

Fig. 4.2 Compartmental toxicokinetic models. **k_a**—*first-order absorption rate constant,* **k_{el}**—*first-order elimination rate constant:* **k_a**—*first-order absorption rate constant into the central compartment (1),* k_{10}—*first-order elimination rate constant from the central compartment (1),* k_{12} and k_{21}—*first-order rate constants for distribution between central (1) and peripheral (2) compartment*

(a) **One-compartment open model (Fig. 4.2)**
It is the most basic model in which the entire body is treated as a single unit.
Intravenous administration: During the IV bolus injection of the test item, the chemical gets disseminated throughout the body through blood circulation to all the tissues. The graph of chemical concentration (drug duration curve) will be a straight line on semi-logarithmic paper, where it display a monophasic decline.

(b) **Two-compartment open model**
The two-compartment model defines that the body consists of two compartments. The central compartment is considered as compartment 1 and the peripheral compartment is considered as compartment 2. When a toxicant or test item is ingested, it takes some time to reach the organs in the body and to get equilibrated with the concentration of body fluids (blood and plasma) in a time-bound manner. The central compartment includes the blood and highly perfused organs like the brain, heart, lung, liver, and kidney while the peripheral compartment is comprised of less perfused tissues like muscle, skin, bone and cartilage.

2. **Non-compartment model:**
In this particular model, the time point of plasma drug concentration as a statistical distribution curve is considered. The kinetic parameters are derived using algebraic equations (Grech et al. 2017).

3. **Physiological-based toxicokinetics (PBTK)**
PBTK models are mathematical simulations of physiologic tissues of organisms, which provide the mechanistic framework for evaluating the absorption, distribution, metabolism and excretion (ADME) of test item or the toxicant in the body. It helps to predict the distribution of toxicants in tissues or organs based on respect to time. The model structure contains the organs or tissues which are essential to describe the important features of ADME of the test item and its metabolites and the target tissues in which toxicity may arise (Mavroudis et al. 2018) (Fig. 4.3).

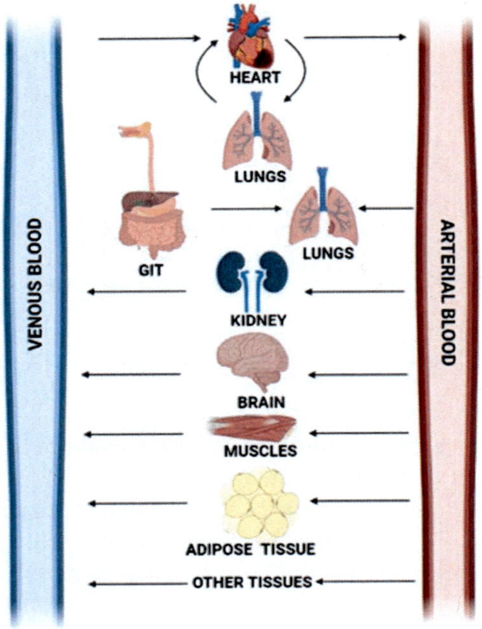

Fig. 4.3 Whole-body physiologically based toxicokinetic model [adapted with slight modification from (Gehring and van der Merwe 2014)]

4.5 Orders of Kinetics

Several rate methods will be encountered during ADME. For example, when a toxicant is given viz. oral or extravascular (EV) route [i.m, s.c, i.p], it gradually enters the circulation and the plasma drug concentration slowly increases to the maximum, where it needs to overcome various biological barriers and metabolism (Gupta 2016)

(a) **Zero-order kinetics**: (constant kinetics)
It is defined as a toxicokinetic method whose rate is independent of the concentration of the test item/chemical, i.e. a constant quantity of chemical is eliminated at a fixed time interval which is independent of chemical concentration (Fig. 4.4)

(b) **First-order kinetics**: (linear kinetics)
It is the toxicokinetic method whose toxicokinetic rate is directly proportional to the chemical/test item concentration (Fig. 4.5)

(c) **Mixed order kinetics**: (nonlinear kinetics)
A toxicokinetic method in which rate is a mixture of both zero and first-order kinetics. It follows zero-order kinetics at higher concentrations and first-order kinetics at a lower concentration of the chemical/test item.

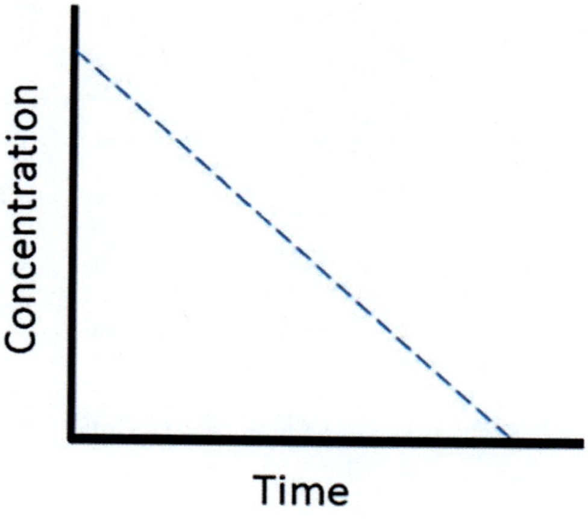

Fig. 4.4 Graph representing concentration versus time for zero-order kinetics reaction

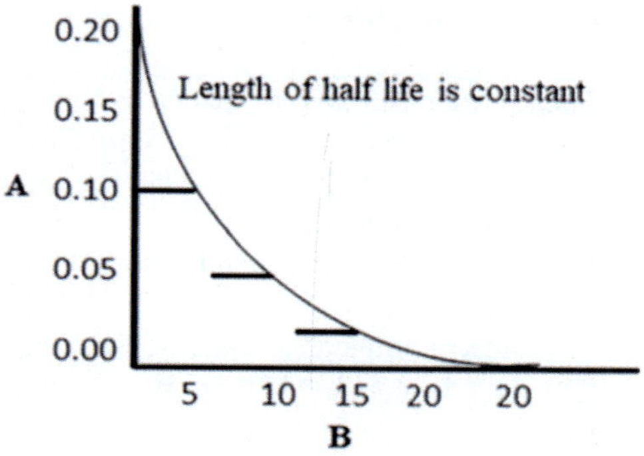

Fig. 4.5 Graph representing concentration versus time for first-order kinetics reaction

(d) **Saturation kinetics**

It is also known as nonlinear toxicokinetics; the rating process of a drug's ADME is dependent upon the carrier or enzymes that are substrate-specific. It has definite capacities and is susceptible to saturation at high drug concentrations.

4.6 Factors Influencing Toxicokinetics

The effects of confounding factors such as pH, temperature, the solubility of the test item, particle size, dosage form, route of administration is integrated into the compartmental TK models. The changes in physiological parameters of an organism also influence the TK of chemicals. More often the compartmental models assume no growth of the organism, which is a reasonable assumption during a short-term laboratory experiment. First-order kinetics is commonly assumed in the majority of compartmental models, i.e. the rate of transfer of a chemical from or to a compartment is directly proportional to the concentration in that compartment. However, methods can also integrate zero, first, second-order, or saturable kinetics.

Procedure (single dose model)		
Species	:	Rodents (preferably rats)
Sex	:	Male and/female (nonpregnant and nulliparous)
Age	:	6–12 weeks (at the time of dosing)
Acclimatization	:	Minimum 5 days prior to the application of dose. Dose: Pilot study: Single oral dose in general. The nontoxic dose should be given and the dose should be high enough to evaluate the metabolite in excreta. Main study: A minimum of two doses is preferred because it helps in setting the dose for other toxicity studies.

4.7 Study State Experiment (Test No. 417 n.d.)

- The test item is mixed or suspended homogeneously using an appropriate vehicle and administered via oral/intravenous route.
- The dosing volume should be kept as lesser as possible for both the aqueous and non-aqueous vehicles.
- For rodents, the dosing volume should not exceed 10 mL/kg body weight.
- If the intravenous route is chosen, the test substance should be mixed or suspended completely and dose–volume should be appropriate (i.e. 1 mL/kg bw). The dosing should be done on at least four animals of each sex.
- Based on the physicochemical properties and proposed human exposure route, the test item is administered in other routes like dermal and inhalational with appropriate justifications.

4.8 Measurement of Parameters

- Mass balance: It is evaluated by the summation of the percentage of concentration excreted in faeces, urine, and expired air, and the percentage present in tissues, cage wash and residual carcass.
- Absorption:
  - It is calculated using the formula below:
    Percent absorption = (amount in bile + urine + expired air + carcass without GI tract contents)/amount administered × 100
- Bioavailability:
  Plasma/blood kinetics of the oral and intravenous group is used to calculate the bioavailability. The formula used is:

$$F = \left(AUC_{exp}/AUC_{IV}\right) \times \left(Dose_{IV}/Dose_{exp}\right)$$

where AUC is the area under the concentration–time curve; exp is the experimental route of administration.

- **Tissue distribution**
  The tissue distribution will provide information on the accumulation of the test item in the biological matrices. Biological matrices include liver, GI tract, fat, spleen, kidney, residual carcass, whole blood, target organ tissues and any other tissues which are important in evaluation.
- **Metabolism**
  The collection of excreta is done to assess the metabolites and the unchanged form of test item. Periodic profiling of metabolites for each dose group is mandatory.
- **Excretion**
  By measuring the test item concentration in faeces, urine and expired air, the elimination rate of the administered dose is determined. If there is no elimination of test item in expired air in the pilot study, then there is no need to collect the same in the definitive study. The collection of urine is done every day, twice after 24 h of drug administration. Excreta collection can be terminated after 7 days or after at least 90% of the prescribed dosage has been expelled (either of the above events which occur earlier).

4.9 Conclusion

Toxicokinetics is one of the most important preclinical investigations in drug discovery and it plays a very crucial role in defining the dose and dosage regimen of the test item. Toxicokinetics provides additional information in toxicity profiling which is used to extrapolate the preclinical data to clinical trials.

References

Baldrick P (2003) Toxicokinetics in preclinical evaluation. Drug Discov Today 8:127–133. https://doi.org/10.1016/s1359-6446(02)02568-0

Casarett LJ, Doull J, Klaassen CD (eds) (2008) Casarett and Doull's toxicology: the basic science of poisons, 7th, 7th edn. McGraw-Hill, New York

Gad SC (n.d.) Preclinical development handbook, p 1347

Gao Y, Zhang Y, Feng J, Zhu L (2019) Toxicokinetic–toxicodynamic modeling of cadmium and lead toxicity to larvae and adult zebrafish. Environ Pollut 251:221–229. https://doi.org/10.1016/j.envpol.2019.05.003

Gehring R, van der Merwe D (2014) Toxicokinetic-toxicodynamic modeling. In: Biomarkers in toxicology. Elsevier, pp 149–153. https://doi.org/10.1016/B978-0-12-404630-6.00008-7

Grech A, Brochot C, Dorne J-L, Quignot N, Bois FY, Beaudouin R (2017) Toxicokinetic models and related tools in environmental risk assessment of chemicals. Sci Total Environ 578:1–15. https://doi.org/10.1016/j.scitotenv.2016.10.146

Gupta PK (2016) Principles and basic concepts of toxicokinetics. In: Fundamentals of toxicology. Elsevier, pp 87–107. https://doi.org/10.1016/B978-0-12-805426-0.00009-3

Mavroudis PD, Hermes HE, Teutonico D, Preuss TG, Schneckener S (2018) Development and validation of a physiology-based model for the prediction of pharmacokinetics/toxicokinetics in rabbits. PLoS One 13:e0194294. https://doi.org/10.1371/journal.pone.0194294

S 3 A Toxicokinetics: A Guidance for Assessing Systemic Exposure in Toxicology Studies (2006) 12

Test No. 417: Toxicokinetics | en | OECD (n.d.). https://www.oecd.org/env/test-no-417-toxicokinetics-9789264070882-en.htm. Accessed 22 July 2021

5 Safety Pharmacology

5.1 Purpose

- To evaluate the undesirable pharmacodynamic effects of the test item that may impose health hazards on humans
- To compare the adverse pharmacodynamic effects of the test item with the toxicology data

5.2 Definition

"Safety pharmacology studies are defined as those studies that investigate the potential undesirable pharmacodynamic effects of a substance on physiological functions in relation to exposure in the therapeutic range and above"—ICH-guideline S7A, 2001. (Anonymous 2018)

The dose selected for the safety pharmacology studies might impart untoward side effects on the targeted and non-targeted organ systems. Safety pharmacology studies utilize electrophysiological, anatomical and surgical, cellular, biochemical and molecular techniques to obtain observational data. The toxicological data along with safety pharmacology data provides more information to the clinical investigators on the potential adverse effects that might occur in the central nervous, cardiovascular and respiratory systems. Hence, safety pharmacology studies need to be performed before the first human exposure, that is clinical trial phase—I (Andrade et al. 2016).

S. B. Chidambaram et al., *Introduction to Toxicological Screening Methods and Good Laboratory Practice*, https://doi.org/10.1007/978-981-16-6092-4_5

5.2.1 ICH Test Guidelines for Non-clinical Safety Studies

ICH S7A	:	Safety pharmacology studies for human pharmaceuticals
ICH S7B	:	Non-clinical evaluation of the potential for delayed ventricular repolarization (QT interval prolongation) by human pharmaceuticals

5.2.2 Safety Pharmacology Tiers

Safety pharmacology studies are performed at two tiers:

Core battery test	:	• Involves investigations on the effects of the test items on the cardiovascular, central nervous and respiratory systems. • The tests are performed initially because an acute adverse effect on the above-mentioned systems could be life-threatening during phase I clinical trial.
Follow-up and supplemental safety pharmacology studies	:	• **Follow-up studies**: To generate greater depth of information in addition to the core battery tests data • **Supplemental studies**: Involve investigation on the effects of the test item on gastrointestinal, renal, autonomic and other organ systems

5.2.3 Major Organ Systems Studied in Safety Pharmacology Are

A. Primary organ systems (Core Battery test) are:
- Cardiovascular system
- Central nervous system
- Respiratory system

B. Secondary organ systems are:
- Gastrointestinal system
- Renal system (Table 5.1)

5.2.4 Follow-Up and Supplemental Safety Pharmacology Studies

Follow-up studies are intended to provide a deeper comprehension of, or additional knowledge that is provided by the core battery on vital functions (Goineau et al. 2013)

- **Central nervous system:** Behaviour, learning and memory, neurochemistry, visual, auditory, ligand-binding assay and/or electrophysiology investigations

Table 5.1 Safety pharmacology core battery of tests (ICH guidelines S7A and S7B)

Core battery test	Parameters
Cardio vascular system	QT interval • hERG channel assay • ECG Cardiac output • Echo-cardiography • Langendorff's isolated heart assay • Left ventricular pressure and contractility Blood pressure • Invasive • Non-invasive Other—clotting time, platelet aggregation etc
Central nervous system	• Functional observation battery (FOB) which includes co-ordination, body temperature, behaviour, neuromuscular, sensory motors, convulsions • Learning and memory • Anxiety and depression
Respiratory system	Non-invasive • Barometric plethysmography • Nose only plethysmography Parameters—rate of respiration, airway resistance, tidal volume, compliance, pCO_2, pO_2, pH etc.
Renal	Diuresis, Urine volume, pH, specific gravity, ketone bodies, electrolytes, total proteins, glomerular filtration rate
Gastrointestinal	Intestinal motility, gastric ulcer, duodenal ulcer, gastric emptying time, acid secretion

- **Cardiovascular system:** Ventricular contractility, vascular resistance, cardiac output, the effects of endogenous and/or exogenous substances on the cardiovascular responses
- **Respiratory system:** Blood gas, pH, airway resistance, pulmonary arterial pressure, compliance and so on (Pugsley et al. 2008)

5.3 Cardiovascular Safety System (Champeroux et al. 2013)

The cardiovascular system (CVS) involves electrical stimuli, ion channels, muscle and vascular functions to maintain hemodynamic homeostasis. The test item should be evaluated for its effects on the blood pressure, heart rate and electrocardiogram (ECG) before the phase-I clinical trial. Cardiovascular safety pharmacology studies are usually conducted in nonhuman primates and dogs.

5.3.1 Blood Pressure and Heart Rate Measurement

Following the administration of a single dose of the test item, the test system should be continuously monitored for arterial blood pressure and heart rate for at least 20 min using invasive or non-invasive blood pressure measurement apparatus.

5.3.1.1 Procedure

On the day of administration, 1-h baseline recordings should be obtained from the non-anaesthetized animals. The test item is administered through the human intended route (oral, intravenous and inhalation) at an appropriate dose and blood pressure is measured at 30, 60 min, 1, 4 and 24 h. The animal is monitored for the development of hemodynamic effects, if any (usually at least 20–30 min as observation period), and further higher doses of the test item can be planned accordingly. It is advisable to withdraw blood samples at the end of each time point of BP measurement for the determination of plasma drug concentration.

5.3.2 Electrocardiogram

ECG tracings are recordings made for 2 min before (baseline), and after 30, 60, 90, 120, 150 and 180 min of test item administration and analysed for QT interval prolongation using ECG analysis software. Based on the RR interval, QT interval correction is established by selecting minimum ten cardiac cycles (Chu et al. 2001).

CVS follow up studies measure the following:

- Cardiac output
- Ventricular contractility
- Vascular resistance

The overall cardiovascular performance is assessed by cardiac output. Cardiac output is the product of heart rate and stroke volume. Heart rate is measured by conventional methods (radial pulse, cardiac pulse, pedal pulse, brachial pulse, etc.), and stroke volume is measured using different imaging techniques like echocardiography, being the most widespread. Ventricular pressure is measured using the maximal rate of pressure development in the left ventricle (LVdP/dtmax) to assess myocardial contractility. In order to calculate vascular resistance, a pressure transducer or a fluid-filled catheter is inserted into the pulmonary artery and vein.

5.3.3 In-Vivo Telemetry Method

Radiotelemetry is a gold standard method used to obtain real-time and reliable hemodynamic data such as BP, ECG, heart rate and so on using a few numbers of animals. It permits for the measurements of hemodynamic parameters in awake and freely moving animals. It involves surgical implantation of telemetry probes (internal and external), which can simultaneously measure arterial, systemic and left ventricular pressure, heart rate and ECG parameters—QRS complex, QT, ST and PR periods (Kramer and Kinter 2003).

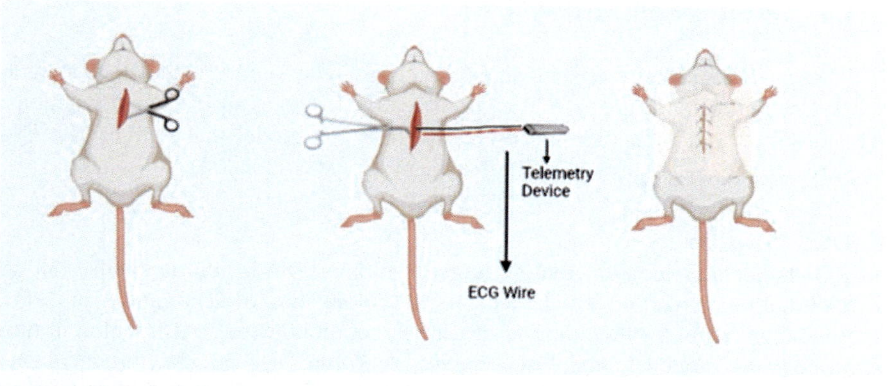

Fig. 5.1 Telemetry unit surgical implantation

5.3.3.1 Procedure

Internal Jacketed

- Animals are anesthetized and the aortic catheter is implanted surgically distal to the renal arteries.
- ECG telemetry unit consists of two electrical ECG primes connected to a radio transmitter.
- The transmitter and battery unit are attached to the abdominal region using a suture.
- One prime is implanted in the right upper chest and the other near the left lower chest.
- A latent period of 1 h is allowed for acclimatization of the animal before test item administration and an initial 15–30 min is also left due to excitement of drug administration.
- The response generated due to the test item leads to the up-regulation of the transmitter; the electrical signals are then transmitted wirelessly to a nearby receiver. The receiver forwards the signal to an amplifier and a computer system is used for data possession, storage and interpretation (Fig. 5.1).

External Jacketed

- Animals are gently positioned on the ECG recording platform.
- An ordered series of gel-lined ECG electrodes are firmly fixed on the floor of the platform and spacing is provided between the electrodes and animal paws. The electrodes are connected to an amplifier.
- After 10 min of acclimatization period, the baseline recording is performed.
- ECG is recorded after 30, 60, 90, 120, 150 and 180 min of test item administration and analysed for QT interval using ECG analysis software
- When the paws are in contact with three electrodes, the signal from the amplifier is accepted or otherwise rejected.
- The data from steady recordings of ECG signals are used for analysis and interpretation.

5.3.4 In-Vitro hERG Assay

The human ether a-go-go-related gene (hERG) is a potassium channel alpha subunit (coded by the KCNH2 gene). It is highly selective for potassium channel functions. The In-vitro evaluation of drug effects on IKr or a model that mimics IKr is a requirement of cardiovascular risk assessment (Priest et al. 2008).

5.3.4.1 Purpose

hERG channels mediate internal currents; disturbance in this current causes slows myocardial repolarization which leads to the prolongation of QT span in ECG. QT prolongation increases the duration of cardiac action potential (AP), which in turn contributes to the development of *Torsades de Pointes*, a fatal arrhythmia. Hence, the effects of a test item on hERG channel function should be evaluated before the first human trial.

5.3.4.2 Procedure

Electrophysiological patch-clamp assays are considered to be the most efficient for ion channel investigations. hERG current is monitored mostly using the voltage-clamp technique in which the transmembrane potential is controlled in the experiment, resulting in ionic current flow across the membrane, and thus the biological response and impact of the test item are recorded (Fig. 5.2).

- In the experiment, non-transfected human embryonic kidney 293 (HEK) cells and transfected HEK cells that expressed wild-type hERG protein (WT-hERG) were used.
- The cells are continuously superfused with extracellular saline solution and pH7.4 is maintained during the entire patch-clamp technique.
- The glass micropipettes for entire patch-clamp recordings are filled with intracellular saline solution mimicking the composition of the cytoplasm.
- The voltage stimulation is depicted below (Fig. 5.3):
- The test item is applied to the cell for 3 min per concentration. The current amplitude value is plotted versus time to visualize the effect of the test item concentration and calculate the percentage of hERG channel inhibition (y)

5.3.4.3 Interpretation

hERG channels are involved in the repolarization phase of cardiac action potentials. Reduced activity of the channels increases the ventricular action potentials, thus prolonging the QT interval in the ECG, which is linked with the fatal *Torsades de Pointes* (ventricular tachyarrhythmia). A drug candidate failing the hERG assay is shown to inhibit the delayed rectifier potassium current. Various HERG assays like flux assays, fluorescence-based assays and binding assays coupled with electrophysiology are employed depending on the test item (Guth 2007).

The percent inhibition is determined using mean data from all recorded cells for each concentration, and the IC50 values are extracted using the Hill equation:

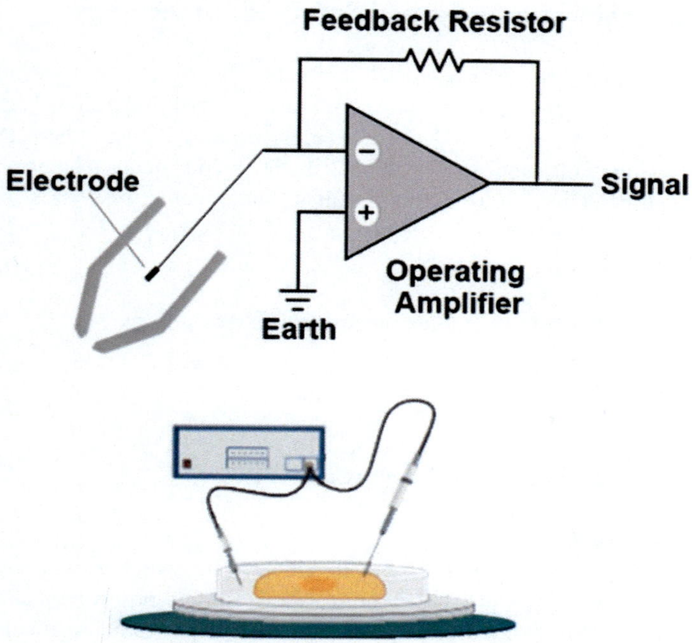

Fig. 5.2 Scheme depicting the generalized recording configurations used in manual and automated electrophysiology recordings

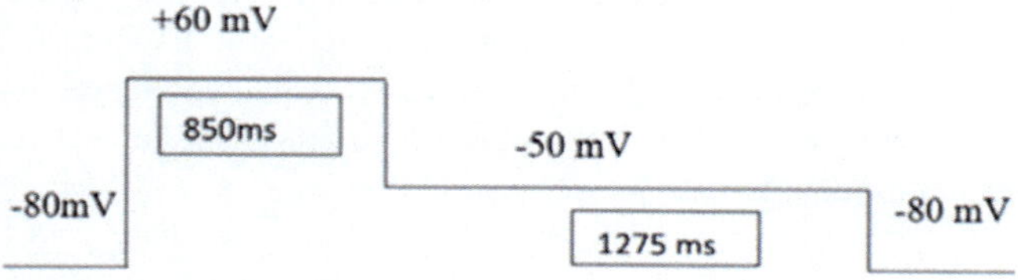

Fig. 5.3 Electrophysiology voltage stimulation

$$y = V_{\max} \frac{x^n}{x^n + k^n}$$

y = mean inhibition rate of each cell; $V_{\max}$ = 100%; x = nominal concentration; n = Hill coefficient; k = test concentration at 50% inhibition.

5.4 Central Nervous System Safety Pharmacology

The potential undesirable pharmacological effects of a test item on CNS function are evaluated by single and repeated dosing studies. In traditional core battery CNS safety pharmacology studies, rodents are widely used to evaluate the safety of the test item using functional observational battery and neurobehavioral assays. Alterations in behavior, motor activity, sensory/motor reflex responses, body temperature and coordination, indicate the adverse effects of a test item on CNS (Castagné et al. 2013).

CNS safety pharmacology studies performed in two phases:

- Core battery tests
- Follow-up studies

5.4.1 Core Battery Tests (Porsolt et al. 2002)

5.4.1.1 Irwin Test

Irwin test is the primary assay in CNS core battery tests, which provides qualitative information on the effects of a test item on behaviour and physiological functions. The outcomes of the test are used to predict the safety of a test item and also help to identify the maximum dose at which no side effects are observed and also an active dose range.

Animals are administered with the test item, behavioural and physiological functions are compared with a control group treated with a vehicle. Physiological and neurotoxicity, behavioural changes, rectal temperature and pupil diameter are recorded as per the standardized protocol mentioned in the Irwin test. The clinical signs are recorded as presence or absence; increased or decreased activity or rated on a 3-point scale. Others parameters like respiration, pupil diameter and rectal temperature are measured quantitatively.

Procedure

Functional Observational Battery (FOB) is performed at various domains as per the Irwin test (Table 5.2).

5.4.2 CNS Follows Up Studies: (Barile 2019)

Based on the previous evidence, the following tests might be considered for follow up studies:

- Behavioural pharmacology
- Learning and memory
- Ligand-specific binding
- Neurochemistry

Table 5.2 Functional observational battery for evaluation of clinical signs

Functional domain	Parameter
Autonomic	1. Pupil diameter 2. Salivation 3. Lacrimation 4. Excessive urination 5. Diarrhea 6. Abnormal respiration 7. Rectal temperature
Neuromuscular	1. Abnormal posture 2. Abnormal gait 3. Catalepsy 4. Body tone 5. Grip strength 6. Tremor 7. Shivering 8. Twitches 9. Convulsions 10. Straub tail 11. Ptosis 12. Exophthalmos
Sensorimotor	1. Approach response 2. Touch response 3. Startle response 4. Palpebral reflex 5. Pinna reflex 6. Tail flick reflex 7. Righting reflex 8. Landing foot splay 9. Grasping reflex 10. Traction response
Behavioural	1. Arousal 2. Sedation 3. Handling reactivity 4. Aggressiveness 5. Spontaneous activity 6. Rearing 7. Sniffing 8. Grooming 9. Scratching 10. Stereotypy 11. Vocalization

- Visual examination
- Auditory examination
- Electrophysiology examinations

5.4.2.1 Behavioural pharmacology

Behavioural pharmacology studies include various tests like sociability, social memory, sexual behaviour assessment and so on.

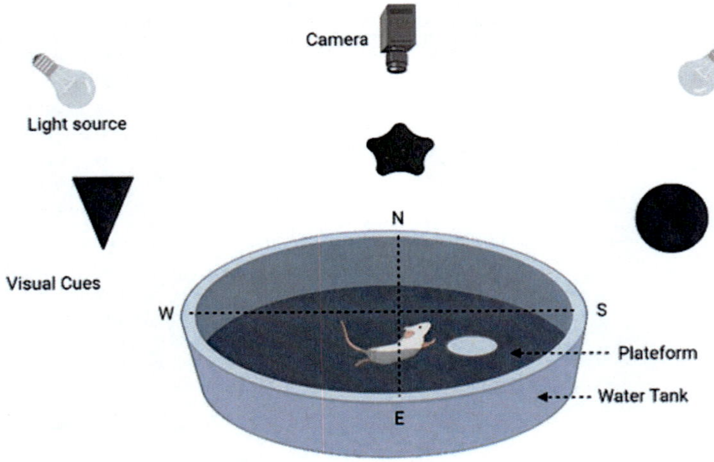

Fig. 5.4 Morris water mazes with different quadrant

Crawley's Sociability test: The main purpose of the test is to assess the social interaction behaviour of the animal, since social interaction is a basic and adaptive aspect of an organism and also indicates the anxiety level. Crawley's sociability apparatus is a rectangular three-chamber box with access from the central chamber to the left and right chambers. The test animal is kept in the central chamber and the stranger animal is placed in the left or right chamber. The wall between the compartments is removed and the test animal is allowed to interact with a stranger animal; the duration and number of contacts between are recorded.

5.4.2.2 Learning and Memory

Assessed using Morris water maze, T mazes, radial arm maze (RAM), Novel objects recognition (NORT) and so on.

1. Morris water maze test

The test is performed following repeated or simultaneous administration of the test item. Morris water maze (MWM) task is a widely used and well-validated test for assessing spatial memory in rodents. It comprises a water-filled circular pool apparatus and a submerged escape platform (1–2 cm) below the water's surface. The rat or mouse learns to locate the platform with the help of visual cues in the room. Further, they are also relocated to a different quadrant or removed during the probe trial experiment. This way the memory retention is investigated. For 5 continuous days, each trial is confined to 2 min in the rat and 1 min in the case of the mouse. Final day, that is on the sixth day, the escape platform is removed and animals are assessed for memory. The time spent in the escape platform quadrant and latency time to reach the platform quadrant is measured using software (Hernier et al. 2016) (Fig. 5.4).

Water: The temperature of the water (20–25 °C).

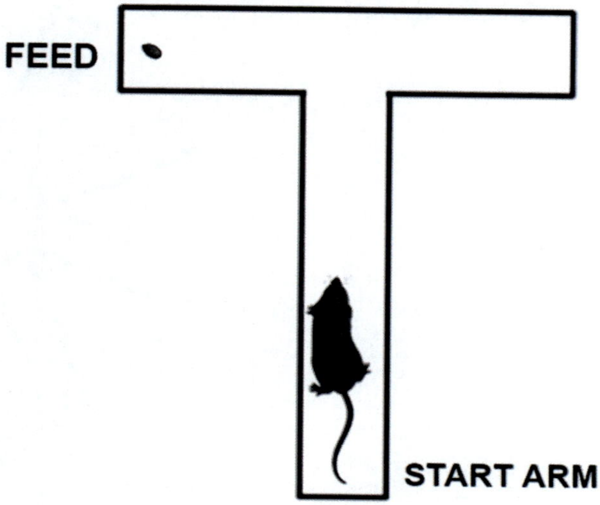

Fig. 5.5 T-maze for rodents (R: Rat and M: Mice)

Room configuration: The maze should be placed in a room with visual cues.
Lighting: ~ 300 Lux

2. T Maze

This test is performed following repeated or simultaneous administration of the test item. T-maze is used to assess spatial memory in rodents. During consecutive trials, animals are trained to distinguish between the two arms using visual, olfactory, tactile and even auditory cues. Following that, memory retention is investigated. Alternatively, the T-maze is sometimes used for place preference testing. Reward within the arms is used to motivate the subject to move to the arms, those can be food or another animal of the same cage, an odour or a type of shelter. Animals are to be kept half an hour before in the testing room to acclimatize. The first step is habituation, animals are left in the maze for about 5 min and during the training animals are rewarded, to learn fast. The trial should be completed in 2 min (120 s). If an animal fails to reach the goal within the 90 s, it should be guided gently. Between each trial, the maze is to be cleaned with 10% Ethanol. Once the animals are trained finally, they are assessed for memory retention (Deacon and Rawlins 2006) (Fig. 5.5).

3. Radial Arm Maze

This test is performed following repeated or simultaneous administration of the test item. The radial arm maze is used to measure spatial, working and reference memory in rodents. When investigating working memory, a reward (usually feed pellet) is placed in all given arms, and the animal should visit the baited arm only once. When investigating reference memory, the bait will be placed only in selected arms and the animal should visit only those baited arms. By relying on cues inside or outside the maze, or by memorizing, the animals learn the location of feed reward present in arms. The test requires 4 consecutive days of training, once animals learn the

Fig. 5.6 Radial arm maze

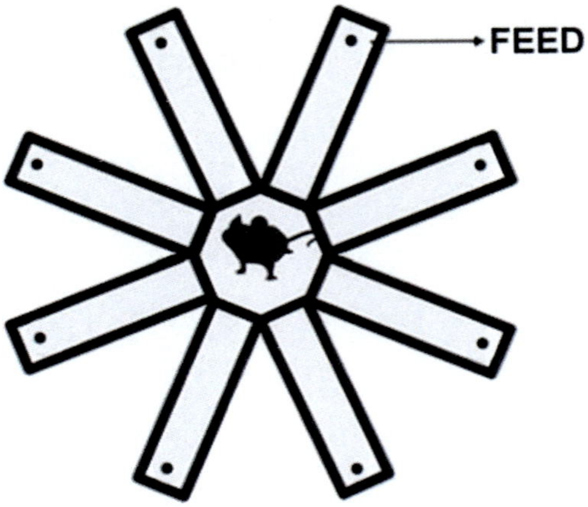

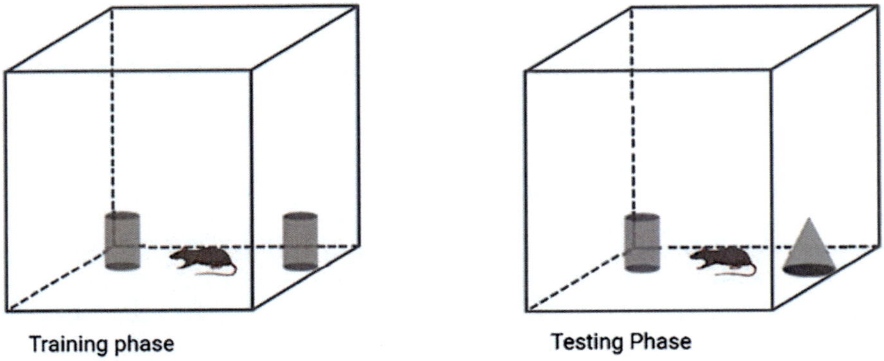

Fig. 5.7 Animals with two similar cues and with one novel cue

navigation in the maze, the final assessment is done with a cut-off time of 5 min (Radial Arm Maze Test - an overview I ScienceDirect Topics n.d.) (Fig. 5.6).

4. Novel Object Recognition

This test is performed following repeated or simultaneous administration of the test item. In an open field box (dimension of the box is different for mouse and rat). During the training phase, the animal is exposed to two familiar objects (5 min), and during the testing phase (generally 4–24 after training), one of the known objects is replaced with a novel object. The exploration time (5 min) is noted for the novel and known objects, which indicate the discrimination ability and recognition memory in animals (Novel Object Recognition 2020) (Fig. 5.7).

Fig. 5.8 Characterizing visual performances in mice

5.4.2.3 Visual Assessment of Rodents

The animal is placed on an elevated platform which is surrounded by a mechanical drum. The inner surface of the drum has vertical gratings. When the drum starts moving, the vestibular input dissociates from the visual input. As a response, the animal reflexively moves its head or the whole body to compensate for the sensory mismatch. The head and body movements are recorded (Benkner et al. 2013; Prusky et al. 2000) (Fig. 5.8).

5.4.2.4 Auditory Function Evaluation

Active electrodes are placed subdermally at the forehead. The reference electrode and ground electrode are placed below the pinna of the left ear and right ear, respectively. The speaker must be placed 10 cm from the left ear. The sounds are presented and ABRs are recorded in a free field condition (Barile 2019) (Fig. 5.9).

5.5 Respiratory System Safety Pharmacology

As per reports, the test drugs induced respiratory abnormalities are low in incidences (0.5–1.2%); however, they are associated with a high mortality index (25–30%) (DiPiro 2005). Similarly, the drug withdrawal from the market due to respiratory adverse events is relatively high. Hence, the data on respiratory safety pharmacology is mandatory before the first exposure in humans. The respiratory system performs two major functions: (1) Ventilation and (2) Gas exchange. The ventilation or pumping function is governed by nerve connections, respiratory muscles, chemoreceptors and mechanoreceptors. The biological structures of gas exchange

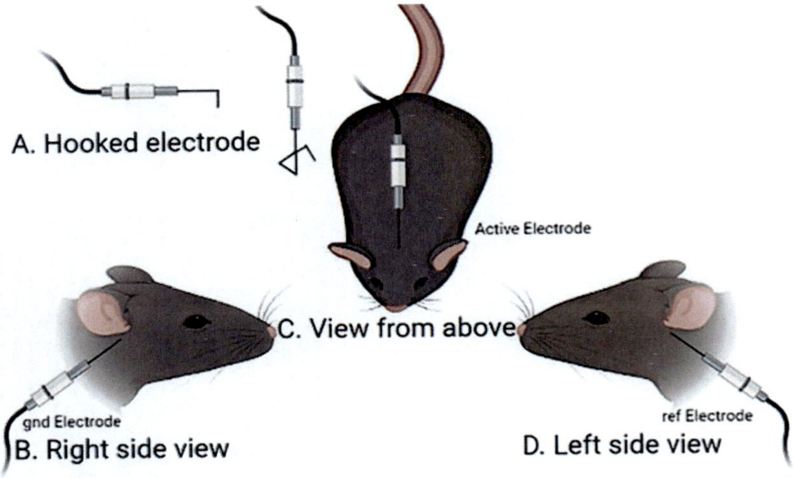

Fig. 5.9 Subdermal implantation of electrodes for recording Auditory Brainstem activity [Adapted with modifications from (Ingham 2019)]

include nasal cavity, trachea, bronchi, bronchioles, alveoli and blood vessels (JaypeeDigital I Drug-induced Pulmonary Diseases n.d.; Murphy 2014). The primary objective of respiratory safety pharmacology is to generate information on the effects of the test items on the above-mentioned two major functions. Respiratory safety pharmacology of the test item is investigated in terms of respiratory rate, airway resistance, compliance, pulmonary arterial pressure, blood gases and blood pH in animal models through invasive and non-invasive methods (Murphy 1994, 2014).

5.5.1 Invasive Methods

5.5.1.1 Whole-Body Plethysmography

The experimental animals are conscious, but unrestrained in whole-body plethysmography and restrained in head out plethysmography. The animals are acclimatized to the plethysmograph chamber for a minimum of 3 days (2 h/day). On the day of the experiment, the basal reading is recorded 30 min after acclimatization. Then, the vehicle or test item is administered and the first reading was taken at 30 min and thereafter, for every 15 min for 1.5 h. The parameters like respiratory rate (RR), tidal volume (TV), mid expiratory flow (EF50), inspiratory time (Ti), expiratory time (Te) peak inspiratory flow (PIF) and peak expiratory flow (PEF) are obtained from the plethysmography data acquisition system. In the invasive body plethysmography system, the animals are anaesthetized and intubated

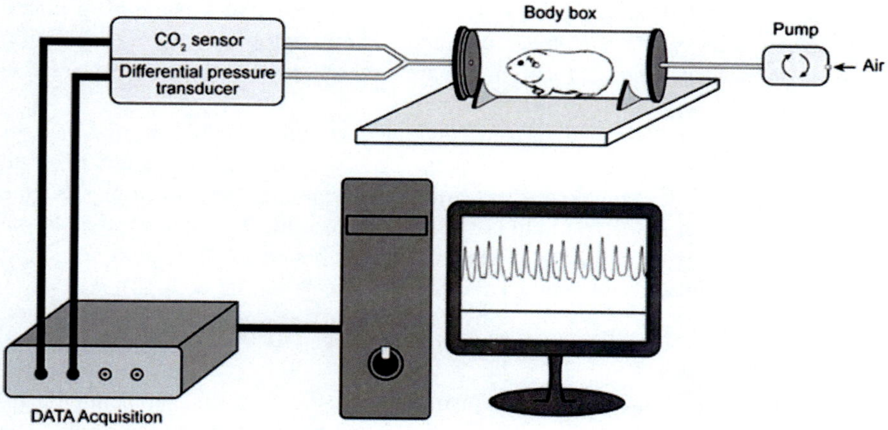

Fig. 5.10 Lung function measurements in rodents in safety pharmacology studies [Adapted with modifications from Pazhoohan et al. 2017]

with a pneumotachograph and recordings are obtained (Hoymann 2012; Murphy 2013) (Fig. 5.10).

5.5.1.2 Blood Profile Analysis

Experimental rats are overnight fasted and anaesthetized using ketamine and xylazine cocktail. The right and left carotid arteries are cannulated for vehicle/test item and for blood sampling (200 μL/time point). The basal sampling was performed prior to the test item administration and then at 30 min, 1 h, 3 h and 5 h post-dose. Blood pH, blood–oxygen tension (pO_2), oxygen saturation (sO_2) and carbon dioxide tension (pCO_2) are analysed using the blood gas analyser.

5.6 Renal System Safety Pharmacology

Drug-induced kidney injury (DIKI) is one of the major causes for the attrition of drugs and withdrawal from the market. The potential of a test item to affect renal function should be evaluated before the first use in humans. Renal safety pharmacology studies are conducted to evaluate the glomerular filtration rate, renal tubular functions and plasma and urine biochemical parameters (Barile 2019; Benjamin et al. 2015).

5.6.1 Assessment of Glomerular Filtration Rate

Glomerular filtration rate (GFR) is a sensitive index of the functional nephron mass. Endogenous small molecules like urea, creatinine, 2-(a-mannopyranosyl)-L-tryptophan (2MPT) and small molecular weight proteins like cystatin C (g-trace), prostaglandin D synthase (b-trace protein), α1-microglobulin, b2-microglobulin serve as biomarkers for renal function and are therefore used to determine the effects of the test item on glomerular filtration rate.

In addition, the clearance tests of endogenous (urea, creatinine, 2-MPT) or exogenous small molecules (inulin, cystatin C, iohexol or iodixanol) are also used to estimate GFR. Clearance testing requires plasma and urine samples as the biological matrices. If the test item is cleared only via the renal system, the plasma clearance is directly proportional to renal clearance; hence, an estimation of the test item in plasma indicates its renal clearance rate (Levey and Inker 2017).

5.6.1.1 Procedure

Following the single or repeated administration of test item at the therapeutic range, the serum or plasma samples are collected from the experimental animals. The biochemical parameters like urea, creatinine, 2-MPTare measured in an automated or semi-automated biochemical analyser using diagnostic kits. The means difference in the level of analysed parameters between the control and test item treated groups is compared to establish the safety pharmacology data. An elevation in the analyte levels outside of the reference range indicates a reduction in GFR and impact of test item.

5.6.2 Evaluation of Tubular Function by Micropunctures

The tubular micropuncture assay is used to evaluate the impact of a test item on the single nephron function in an intact kidney (Fig. 5.11).

5.6.2.1 Procedure

The overnight fasted rats are anaesthetized and the body temperature is maintained using a thermostatically heated table. The femoral artery is cannulated for monitoring the blood pressure and the test item is administered through the human intended route.

An incision is made in the flank to expose the left kidney, which is immersed in mineral oil (37 °C). A stereomicroscope is used to focus the proximal tubules in the kidney surface, and a micropipette (glass capillary tube with an exterior diameter of 8–10 mm) is inserted. A small volume of mineral oil is infused to block the retrograde flow of the perfusion fluid and the cannulated tubules are infused with a balanced electrolyte solution containing inulin (fluorescein isothiocyanate conjugated) and a dye solution (lissamine green). The distal collection is measured by the dye content from the perfused segment. The second pipette with an oil block is inserted and the subsequent infusion fluid is collected for analysis. The mean difference in inulin concentration between the vehicle and test item administration is compared (Micropuncturing the Nephron n.d.; Quamme and Dirks 1986).

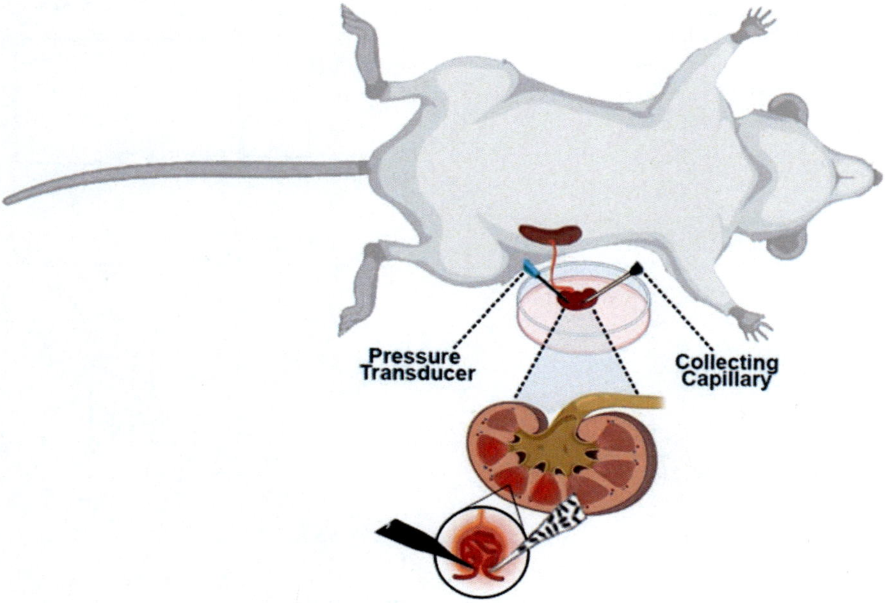

Fig. 5.11 Experimental set up of the multiphoton-based real-time kidney infection model

5.6.3 Diuretic Activity in Rats (Lipschitz Test)

A simple method for testing diuretic activity of test items in the rat model was described by Lipschitz et al. (1943) and Kau et al. (1984). The test is based on sodium and water excretion in animals treated with test item along with a high dose of urea (Fig. 5.12).

5.6.3.1 Procedure

Rats are commonly used for diuretic screening test. The animals are housed individually in the metabolic cages; food and water are withheld for at least 18 h prior to the experiment. The test item is given via the human intended route. The reference control is administered with urea (1 g/kg) in identical dose volume. The negative control group is administered with the vehicle used for test item/reference item preparation. All the animals are hydrated with 25 mL/kg, p.o. of water, 20–30 min after vehicle or test item administration. Then, the animals are kept in the metabolic cage and a urine sample is collected for 5 h/24 h. Urine volume, pH and electrolytes (Na^+, K^+ and Cl^-) are measured using the urine analyser (Vogel 2007).

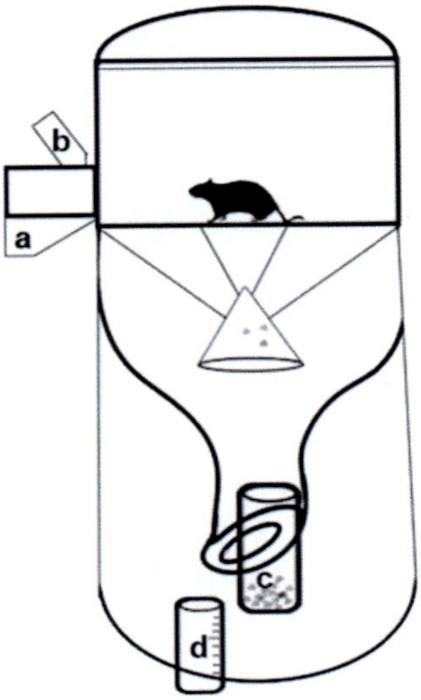

Fig. 5.12 Metabolic cage including containers for (**a**) food (**b**) water (**c**) faeces (**d**) urine [Adapted from Bari et al. 2020]

5.7 Gastrointestinal (GI) Safety Pharmacology

Effects of the test item on GI system functions viz. gastric secretion, pH, mucosal damage potential, bile secretion, intestinal transit time and ileal contraction are evaluated to establish the safety pharmacology data, which in turn improves the prediction of clinical effects. The majority of functions of the gastrointestinal tract (GIT) are similar across species; however, there are significant differences in human and laboratory animals, anatomy, physiology and biochemistry, which are also prioritized while conducting a GI assessment (Pugsley et al. 2008).

5.7.1 Gastric Acid Secretion in Anesthetized Stomach-Lumen Perfused Rats

The continuous recording technique of gastric acid secretion was first proposed by Ghosh and Schild (1958) in the stomach-lumen perfused anesthetized rat model. In this model, gastric acid secretion is stimulated using histamine, gastrin or carbachol. The test item is characterized for its gastric acid anti-secretory potential during the stimulated gastric acid secretion (Herling 2014; Jennewein 1977) (Fig. 5.13).

5.7.1.1 Procedure

The overnight fasted rats (weighing 300–350 g) are anesthetized using ketamine (40 mg/kg) and xylazine (6 mg/kg) plus subcutaneous injection of ketamine as a maintenance dose. The trachea is exposed and cannulated for artificial respiration and body temperature is maintained using a heating pad. The jugular vein is exposed

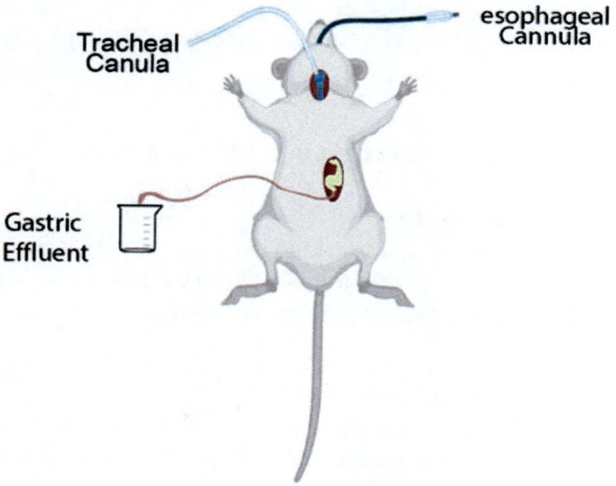

Fig. 5.13 Perfusion of rat stomach and collection of gastric effluent. [Adapted from Nwaichi et al. 2013]

and cannulated with bevelled polyethylene tubes at the tip for injection of secretion stimulating agents. A midline incision is used to open the abdominal cavity. Oesophageal and pyloric ends are ligated and a double-lumen perfusion cannula is inserted and fixed in the fore-stomach. The stomach is perfused continuously with warm saline (37 °C). The perfusate is collected every 15 min and the acid concentration is determined. Histamine (10 mg/kg/h), carbachol (30 mg/kg/h) or desglugastrin (100 mg/kg/h) is administered via the jugular vein after a basal recording a period of 45 min. The test item is injected intravenously 90 min after the secretagogue administration and when the acid output reaches a stable plateau. The gastric perfusate is collected at 15-min intervals. The acid content is determined by titration against NaOH to an endpoint of pH 7 and the acid output (mmol H^+/ 15 min) is calculated (Harrison et al. 2004; Herling 2014).

5.7.2 Effect of the Test Item on Serum Gastrin Levels

When the total acid secretion is blocked (e.g. with omeprazole, an H+/K + -ATPase inhibitor), a gastric antral feedback mechanism is triggered, resulting in severe hypergastrinemia, which on chronic period causes diffuse endocrine cell hyperplasia, characterized as carcinoids, in the gastric corpus in the rat (Al-Saffar et al. 2015; Stomach Acid Secretion - an overview | ScienceDirect Topics n.d.).

5.7.2.1 Procedure

Rats (150–180 g) are pre-treated with the test item and omeprasole as standard drug. The blood sampling is performed on the 2nd, 4th, 7th and 10th week following the exposure to the test item via retro-orbital puncture. Gastrin content is measured by using the ELISA or radioimmunoassay kit. Serum gastrin level is expressed as pg/mL. At the end of the 10 weeks study, the animals are evaluated for gastric acid output using the pylorus ligation method (Shay operated model) (Stroff et al. 1995).

5.7.3 Bile Secretion in a Mouse Model

5.7.3.1 Purpose

The potential of a test item to decrease the bile secretion predominantly represents safety issues in human use. The effect of the test item on bile secretion is assessed in mice by weighing the gall bladder filled with bile.

5.7.3.2 Procedure

Mice (20–25 g) are deprived of feed for 4–6 h with free access to water. The test item is given via the human intended route. The blood sample is collected 1 h after the test item administration. Laparotomy is performed to expose the liver, and a 0.75 silk ligature is placed around the cystic duct, which is separated from the bile ducts and removed from the peritoneal cavity. If the volume of bile is excessive, the gall bladder and bile ducts are removed together. The contents of the isolated gall bladder are collected and weighed. The gall bladder wall is washed with clean water and blotted using filter paper, and the organ is weighed. The quantity of bile secreted is determined by the difference in weight between the full and empty gall bladder (Groen et al. 1995). The concentration of cholate, bilirubin and cholesterol in the bile is determined and compared with the vehicle-treated control mice.

5.7.4 Bile Duct and Duodenum Cannulation Method

The experimental animals are anaesthetized, using ketamine (40 mg/kg) and xylazine (6 mg/kg) plus subcutaneous injection of ketamine as a maintenance dose. The body temperature is maintained using a heating pad. Laparotomy is performed to expose the bile duct and duodenum, and the common bile duct is ligated at 1 cm distal to the hepatic duct junction and the second ligature is placed at 0.5 cm proximal to the duodenum. The bile duct is now incised and a polyurethane cannula is inserted (Fig. 5.14). The cannula (0.7 mm i.d., 1.0 mm o.d.) contains a suture bead 1 cm from the tip and a bevel formed at the tip after being stretched. A suture is tied to keep the cannulation in place and care should be taken to avoid overtightening and therefore restrict bile flow. The cannula's free end is then tunneled with a trochar till it exits at the ventral tail, approximately 20 mm from its base (Kunne et al. 2013).

The duodenum is cannulated approximately 1 cm proximal to the Oddi sphincter. The suture bead and the tip of this cannula are inserted into the duodenum and held in place with a purse-string suture. After that, the duodenal cannula is tunneled to the tail. The abdominal wall is then closed and the skin sutured (Bodewes et al. 2015).

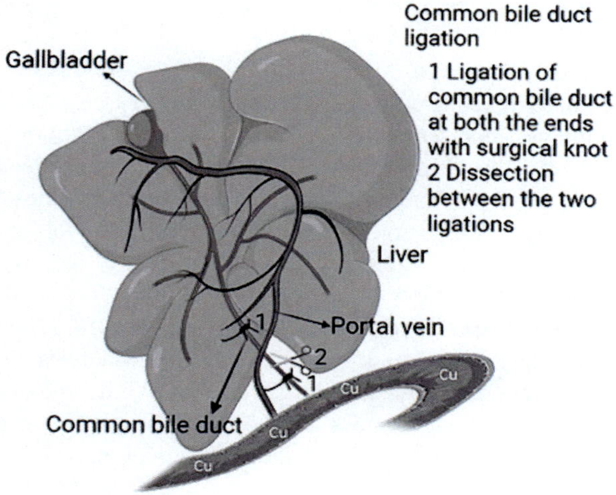

Fig. 5.14 Site of ligation and cannulation of the bile duct

5.7.5 Exocrine Pancreatic Secretion in Anesthetized Rats

5.7.5.1 Purpose

From the GI safety pharmacological assessment point of view, the potential of a test item to decrease exocrine pancreatic secretion is harmful, because it induces pancreatitis. The effect of the test item on pancreatic secretion is measured in rats with acute pancreatic fistula (Shiratori et al. 1986).

5.7.5.2 Procedure

Overnight fasted rat (150–200 g) is anaesthetized using ketamine (40 mg/kg) and xylazine (6 mg/kg) plus subcutaneous injection of ketamine as the maintenance dose. The body temperature is maintained using a heating pad. To ensure uninterrupted breathing, the trachea is cannulated. A midline incision is used to open the abdomen, and the pyloric end is ligated. Near the hepatic portal, the bile duct's proximal end is ligated. The bile secretion is emptied into the duodenum through a thin polyethylene tube (Sata et al. 1996). Another thin polyethylene tube is used to cannulate the distal segment of the bile duct as well as the orifices of pancreatic ducts. After a pre-test period of 60 min, the test item is administered. At 15, 30 and 60 min following test item administration, the pancreatic juice is collected in an Eppendorf tube, and the secreted volume is determined gravimetrically (Pedersen et al. 2000). The bile secretion pre and post-test item administration is compared. Secretin or cholecystokinin (CCK), used to increase pancreatic secretion serves as a reference compound (Petersen and Grossman 1977).

5.7.6 Gastrointestinal Injury

5.7.6.1 Purpose

To determine the potential of a test item to cause gastric mucosal injury. The positive findings represent serious safety issues in human use. Nonsteroidal anti-inflammatory drugs (NSAIDs), like aspirin, indomethacin, ibuprofen inhibits cyclo-oxygenase (COX) enzymes which lead to decreased production of prostacyclin in turn gastric ulceration (Adinortey et al. 2013; Woolf and Rose 2021).

5.7.6.2 Procedure

The experimental rats (150–200 g) are fasted for 24 h and the test item was administrated orally. The rats are sacrificed 6 h later and the stomach, and intestines are dissected out (Simões et al. 2019). The mucosa is examined under a stereomicroscope and the ulceration index was scored. The ulcer scoring is performed as mentioned below:

0	:	No ulcer
1	:	Superficial ulcer
2	:	Deep ulcers
3	:	Perforation

5.7.7 Gastric Ulceration: Pylorus Ligation Model in Rats

5.7.7.1 Purpose

Shay-operated rat model is widely used to determine the gastric ulceration potential of the test item (Monteiro et al. 2013).

5.7.7.2 Procedure

The rats (150–180 g) should be fasted for 48 h, but had free access to drinking water. The animals are housed individually in cages with raised wire meshed bottom to avoid coprophagy. The animal is anesthetized and a midline abdominal incision is performed (Bhajoni et al. 2016). The pyloric end is identified and ligated with caution to avoid damaging the blood supply and the abdominal cavity is closed with sutures (Fig. 5.15). The test item is administered via the human intended route. After 6 h, the animals are sacrificed, the abdomen is opened and a ligature is placed around the oesophageal end. The contents of the stomach are drained into a centrifuge tube. The stomach is opened along the greater curvature and the mucosa is examined under the microscope (Mekonnen et al. 2020; Sakat et al. 2012). Ulceration is scored as mentioned above.

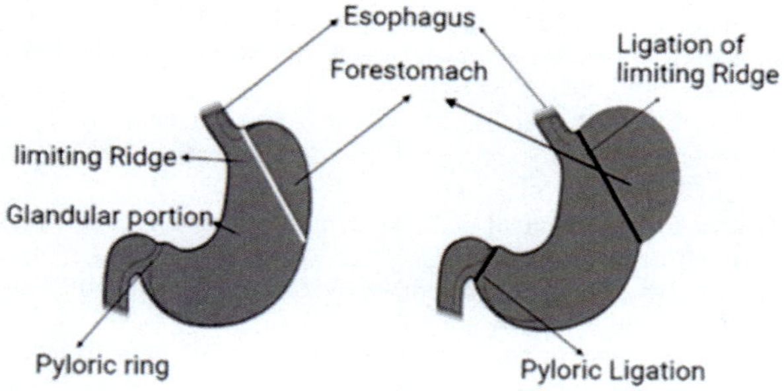

Fig. 5.15 Site of Piloric ligation

References

Adinortey MB, Ansah C, Galyuon I, Nyarko A (2013) In vivo models used for evaluation of potential antigastroduodenal ulcer agents. Ulcers 2013:e796405. https://doi.org/10.1155/2013/796405

Al-Saffar A, da Costa AN, Delaunois A, Leishman DJ, Marks L, Rosseels M-L, Valentin J-P (2015) Gastrointestinal safety pharmacology in drug discovery and development. In: Pugsley MK, Curtis MJ (eds) Principles of safety pharmacology, handbook of experimental pharmacology. Springer, Berlin Heidelberg, Berlin, Heidelberg, pp 291–321. https://doi.org/10.1007/978-3-662-46943-9_12

Andrade EL, Bento AF, Cavalli J, Oliveira SK, Schwanke RC, Siqueira JM, Freitas CS, Marcon R, Calixto JB (2016) Non-clinical studies in the process of new drug development—Part II: Good laboratory practice, metabolism, pharmacokinetics, safety and dose translation to clinical studies. Braz J Med Biol Res 49:e5646. https://doi.org/10.1590/1414-431X20165646

Anonymous (2018) ICH S7A Safety pharmacology studies for human pharmaceuticals. European Medicines Agency. https://www.ema.europa.eu/en/ich-s7a-safety-pharmacology-studies-human-pharmaceuticals. Accessed 23 July 2021

Bari IN et al (2020) Effect of Plumeria rubra (Apocynaceae) leaf extracts, a repellent of rice-field rats (Rattus argentiventer), on its metabolism and daily activity. Rev Biol Trop 68(4):1357–1370

Barile FA (2019) Principles of toxicology testing. CRC Press

Benjamin A, Gallacher DJ, Greiter-Wilke A, Guillon J-M, Kasai C, Ledieu D, Levesque P, Prelle K, Ratcliffe S, Sannajust F, Valentin J-P (2015) Renal studies in safety pharmacology and toxicology: a survey conducted in the top 15 pharmaceutical companies. J Pharmacol Toxicol Methods 75:101–110. https://doi.org/10.1016/j.vascn.2015.01.004

Benkner B, Mutter M, Ecke G, Münch TA (2013) Characterizing visual performance in mice: an objective and automated system based on the optokinetic reflex. Behav Neurosci 127:788–796. https://doi.org/10.1037/a0033944

Bhajoni PS, Meshram GG, Lahkar M (2016) Evaluation of the antiulcer activity of the leaves of azadirachtaindica: an experimental study. IMI 3:10–16. https://doi.org/10.1159/000442750

Bodewes FAJA, Bijvelds MJ, de Vries W, Baller JFW, Gouw ASH, de Jonge HR, Verkade HJ (2015) Cholic acid induces a CFTR dependent biliary secretion and liver growth response in mice. PLoS One 10:e0117599. https://doi.org/10.1371/journal.pone.0117599

Castagné V, Froger-Colléaux C, Esneault E, Marie HA, Lemaire M, Porsolt RD (2013) Central Nervous System (CNS) safety pharmacology studies. In: Vogel HG, Maas J, Hock FJ, Mayer D (eds) Drug discovery and evaluation: safety and pharmacokinetic assays. Springer, Berlin, Heidelberg, pp 17–72. https://doi.org/10.1007/978-3-642-25240-2_3

Champeroux P, Guth BD, Markert M, Rast G (2013) Methods in cardiovascular safety pharmacology. In: Vogel HG, Maas J, Hock FJ, Mayer D (eds) Drug discovery and evaluation: safety and pharmacokinetic assays. Springer, Berlin, Heidelberg, pp 73–97. https://doi.org/10.1007/978-3-642-25240-2_4

Chu V, Otero JM, Lopez O, Morgan JP, Amende I, Hampton TG (2001) Method for non-invasively recording electrocardiograms in conscious mice. BMC Physiol 1:6. https://doi.org/10.1186/1472-6793-1-6

Deacon RMJ, Rawlins JNP (2006) T-maze alternation in the rodent. Nat Protoc 1:7–12. https://doi.org/10.1038/nprot.2006.2

DiPiro JT (2005) Pharmacotherapy: a pathophysiologic approach. McGraw-Hill Medical, New York

Ghosh MN, Schild HO (1958) Continuous recording of acid gastric secretion in the rat. Br J Pharmacol Chemother 13:54–61

Goineau S, Lemaire M, Froget G (2013). Overview of safety pharmacology. Curr Protoc Pharmacol 63, Unit 10.1. https://doi.org/10.1002/0471141755.ph1001s63

Groen AK, van Wijland MJA, Frederiks WM, Smit JJM, Schinkel AH, Oude Elferink RPJ (1995) Regulation of protein secretion into bile: studies in mice with a disrupted mdr2 p-glycoprotein gene. Gastroenterology 109:1997–2006. https://doi.org/10.1016/0016-5085(95)90768-8

Guth BD (2007) Preclinical cardiovascular risk assessment in modern drug development. Toxicol Sci 97:4–20. https://doi.org/10.1093/toxsci/kfm026

Harrison AP, Erlwanger KH, Elbrønd VS, Andersen NK, Unmack MA (2004) Gastrointestinal-tract models and techniques for use in safety pharmacology. J Pharmacol Toxicol Methods 49:187–199. https://doi.org/10.1016/j.vascn.2004.02.008

Herling AW (2014) Pharmacological effects on gastric function. In: Hock FJ (ed) Drug discovery and evaluation: pharmacological assays. Springer, Berlin Heidelberg, Berlin, Heidelberg, pp 1–87. https://doi.org/10.1007/978-3-642-27728-3_56-1

Hernier AM, Froger-Colléaux C, Castagné V (2016) CNS safety pharmacology: a focus on cognitive functions. J Pharmacol Toxicol Methods 81:286–294. https://doi.org/10.1016/j.vascn.2016.04.002

Hoymann HG (2012) Lung function measurements in rodents in safety pharmacology studies. Front Pharmacol 3:156. https://doi.org/10.3389/fphar.2012.00156

Ingham NJ (2019) Evoked potential recordings of auditory brainstem activity in the mouse: an optimized method for the assessment of hearing function of mice. Bio-protocol 9:e3447–e3447

JaypeeDigital | Drug-induced Pulmonary Diseases (n.d.). https://ortholibrary.in/book/9789351520054/chapter/ch4. Accessed 24 July 2021

Jennewein HM (1977) The effect of clonidine on gastric acid secretion in rats and dogs. Naunyn Schmiedeberg's Arch Pharmacol 297:85–90. https://doi.org/10.1007/BF00508814

Kau ST, Keddie JR, Andrews D (1984) A method for screening diuretic agents in the rat. J Pharmacol Methods 11:67–75. https://doi.org/10.1016/0160-5402(84)90054-8

Kramer K, Kinter LB (2003) Evaluation and applications of radiotelemetry in small laboratory animals. Physiol Genomics 13:197–205. https://doi.org/10.1152/physiolgenomics.00164.2002

Kunne C, Acco A, Hohenester S, Duijst S, de Waart DR, Zamanbin A, Elferink RPJO (2013) Defective bile salt biosynthesis and hydroxylation in mice with reduced cytochrome P450 activity. Hepatology 57:1509–1517. https://doi.org/10.1002/hep.26133

Levey A, Inker L (2017) Assessment of glomerular filtration rate in health and disease: a state of the art review: state of the art review for clinical pharmacology and therapeutics. Clin Pharmacol Ther 102:405–419. https://doi.org/10.1002/cpt.729

Lipschitz WL, Hadidian Z, Kerpcsar A (1943) Bioassay of Diuretics. J Pharmacol Exp Ther 79:97–110

Mekonnen AN, Asrade Atnafie S, Wahab Atta MA (2020) Evaluation of antiulcer activity of 80% methanol extract and solvent fractions of the root of croton macrostachyus Hocsht: Ex Del. (Euphorbiaceae) in rodents. Evid Based Complement Alternat Med 2020:e2809270. https://doi.org/10.1155/2020/2809270

Micropuncturing the Nephron (n.d.). https://www.ncbi.nlm.nih.gov/pmc/articles/PMC2954491/. Accessed 24 July 2021

Monteiro KM, Spindola HM, Possenti A, Tinti SV, Ruiz ALTG, Longato GB, Fiorito GF, Marchetti GM, Shiozawa L, Piloni BU, de Oliveira AC, Miyagawa LM, Carvalho JE (2013) Characterization of a refinement of the "pylorus ligation" model of rat gastric ulceration resulting in "no pain" and a more specific pharmacological response. J Pharmacol Toxicol Methods 67:121–128. https://doi.org/10.1016/j.vascn.2012.09.002

Murphy DJ (1994) Safety pharmacology of the respiratory system: techniques and study design. Drug Dev Res 32:237–246. https://doi.org/10.1002/ddr.430320408

Murphy DJ (2013) Respiratory function assays in safety pharmacology. In: Vogel HG, Maas J, Hock FJ, Mayer D (eds) Drug discovery and evaluation: safety and pharmacokinetic assays. Springer, Berlin, Heidelberg, pp 185–197. https://doi.org/10.1007/978-3-642-25240-2_6

Murphy DJ (2014) Respiratory safety pharmacology—current practice and future directions. Regul Toxicol Pharmacol 69:135–140. https://doi.org/10.1016/j.yrtph.2013.11.010

Novel Object Recognition (2020) Histogenotech. https://histogene.co/novel-object-recognition-test/. Accessed 23 July 2021

Nwaichi E, Gwotmut M, Ossai J (2013) Effects of ranitidine on insulin and lime—induced gastric secretion in albinowistar rats. Int J Clin Pharmacol Toxicol 2:38–44. https://doi.org/10.19070/2167-910X-130008

Pazhoohan S, Raoufy MR, Javan M, Hajizadeh S (2017) Effect of Rho-kinase inhibition on complexity of breathing pattern in a Guinea pig model of asthma. PLoS One 12:e0187249. https://doi.org/10.1371/journal.pone.0187249

Pedersen NL-R, Nagain-Domaine C, Mahé S, Chariot J, Rozé C, Tomé D (2000) Caseinomacropeptide specifically stimulates exocrine pancreatic secretion in the anesthetized rat. Peptides 21:1527–1535. https://doi.org/10.1016/S0196-9781(00)00307-7

Petersen H, Grossman MI (1977) Pancreatic exocrine secretion in anesthetized and conscious rats. Am J Physiol Endocrinol Metab 233:E530. https://doi.org/10.1152/ajpendo.1977.233.6.E530

Porsolt RD, Lemaire M, Durmuller N, Roux S (2002) New perspectives in CNS safety pharmacology. Fundam Clin Pharmacol 16:197–207. https://doi.org/10.1046/j.1472-8206.2002.00061.x

Priest BT, Bell IM, Garcia ML (2008) Role of hERG potassium channel assays in drug development. Channels (Austin) 2:87–93. https://doi.org/10.4161/chan.2.2.6004

Prusky GT, West PWR, Douglas RM (2000) Behavioral assessment of visual acuity in mice and rats. Vis Res 40:2201–2209. https://doi.org/10.1016/S0042-6989(00)00081-X

Pugsley MK, Authier S, Curtis MJ (2008) Principles of safety pharmacology. Br J Pharmacol 154:1382–1399. https://doi.org/10.1038/bjp.2008.280

Quamme GA, Dirks JH (1986) Micropuncture techniques. Kidney Int 30:152–165. https://doi.org/10.1038/ki.1986.167

Radial Arm Maze Test - an overview | ScienceDirect Topics (n.d.). https://www.sciencedirect.com/topics/biochemistry-genetics-and-molecular-biology/radial-arm-maze-test. Accessed 23 July 2021

Sakat SS, Tupe P, Juvekar A (2012) Gastroprotective effect of Oxalis corniculata (whole plant) on experimentally induced gastric ulceration in Wistar rats. Indian J Pharm Sci 74:48. https://doi.org/10.4103/0250-474X.102543

Sata N, Kimura W, Muto T, Mineo C (1996) Exocrine function of caerulein-induced acute pancreatitis in anesthetized rats. J Gastroenterol 31:94–99. https://doi.org/10.1007/BF01211193

Shiratori K, Chen YF, Chey WY, Lee KY, Chang T-M (1986) Mechanism of increased exocrine pancreatic secretion in pancreatic juice-diverted rats. Gastroenterology 91:1171–1178. https://doi.org/10.1016/S0016-5085(86)80013-0

Simões S, Lopes R, Campos MCD, Marruz MJ, da Cruz MEM, Corvo L (2019) Animal models of acute gastric mucosal injury: macroscopic and microscopic evaluation. Animal Model Exp Med 2:121–126. https://doi.org/10.1002/ame2.12060

Stomach Acid Secretion - an overview | ScienceDirect Topics (n.d.). https://www.sciencedirect.com/topics/biochemistry-genetics-and-molecular-biology/stomach-acid-secretion. Accessed 24 July 2021

Stroff T, Plate S, Respondek M, Müller K-M, Peskar BM (1995) Protection by gastrin in the rat stomach involves afferent neurons, calcitonin gene-related peptide, and nitric oxide. Gastroenterology 109:89–97. https://doi.org/10.1016/0016-5085(95)90272-4

Vogel HG (2007) Activity on urinary tract. In: Vogel HG (ed) Drug discovery and evaluation. Springer, Berlin Heidelberg, Berlin, Heidelberg, pp 457–510. https://doi.org/10.1007/978-3-540-70995-4_4

Woolf A, Rose R (2021) Gastric Ulcer. In: StatPearls. StatPearls Publishing, Treasure Island, FL

Good Laboratory Practice

6

6.1 Definition

Good Laboratory Practice (GLP) is defined as 'a quality system concerned with the organisational process and the conditions under which non-clinical health and environmental safety studies are planned, performed, monitored, recorded, archived and reported'—OECD (OECD GLP n.d.)

6.2 History

In the early 1970s, US-FDA identified fraudulent animal toxicological reports on drugs, pesticides, food additives, cosmetics and cleaning materials submitted by more than 40 labs, including Industrial Bio-Test Laboratories, across the country. The investigation by US-FDA identified various testing lacunae such as poor-quality control programs, lack of trained manpower, absence of study protocols or major deviations from the protocols, poor characterisation of test items, lack of industrial standards in toxicological testing, deliberate preparation of fraudulent records and reports and administrative problems (Baldeshwiler 2003). The above-mentioned issues led to the implementation of new federal laws for chemical and pharmaceutical companies and also the inspection of laboratories involved in animal research. In 1976, US-FDA proposed the GLP regulations to ensure the quality and integrity of non-clinical laboratory studies. The final GLP regulations were published on 22nd Dec 1978 and enforced as law on 20th June 1979. The OECD governs the GLP program and strives to develop high-quality and reliable non-clinical tests data used for determining the safety of chemicals. Also, OECD continuously reviews and revises the methodologies to ensure safer data.

S. B. Chidambaram et al., *Introduction to Toxicological Screening Methods and Good Laboratory Practice*, https://doi.org/10.1007/978-981-16-6092-4_6

Fundamental objectives of GLP principles are:

- To ensure high quality of data with respect to safety studies
- To ensure auditable and comparable data
- To promote mutual recognition of data and thereby, to limit the wastage of resources

6.3 OECD Principles of Good Laboratory Practice

- The main goal is to help the scientists obtain results which are:
  - Reliable
  - Repeatable
  - Auditable
  - Recognised by scientists worldwide
- To set guidelines for the non-clinical safety studies and to review and update the same
- To set organisational requirements for labs involved in safety studies
- To set documentation procedures to be followed in safety studies
- To make the incidence of false-negatives (e.g. results demonstrating non-toxicity of a toxic substance) more obvious
- To hold legal binding on the safety studies and action against fraudulent data
- To promote mutual acceptance of study data across the international frontiers

6.4 Scope of OECD GLP Principles

- Test items that require safety studies for the registration have to be subjected to regulatory non-clinical health and environmental safety studies following OECD Principles on GLP
- The major class of test items includes
  - Pharmaceuticals (human and veterinary drugs)
  - Agrochemicals (pesticides, herbicides, etc)
  - Food and feed additives
  - Cosmetic products
  - Industrial chemicals

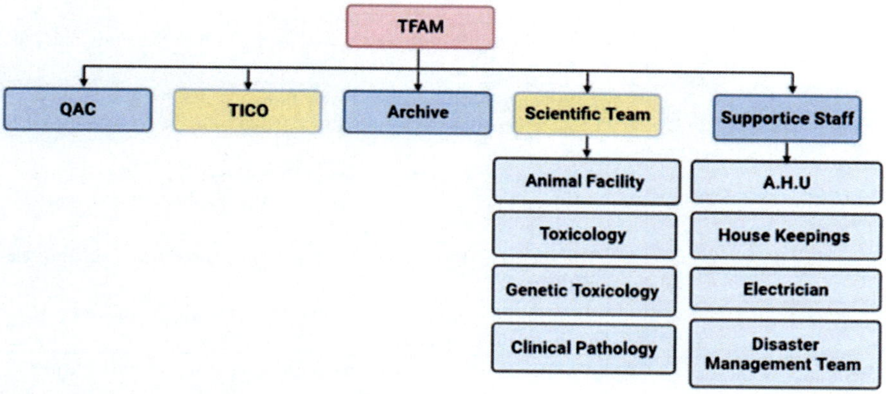

Fig. 6.1 General GLP organogram. *TFM* test facility management; *QAU* quality assurance unit; *TICO* test item control office; *AHU* air handling unit

6.5 National GLP Compliance Monitoring Authority (NGCMA), Government of India

National GLP Compliance Monitoring Authority (NGCMA) was established by the Department of Science & Technology (DST), Government of India, in 2002. In India, the industries, test facilities and laboratories involving non-clinical safety studies are audited and certified for GLP by NGCMA, GoI.

NGCMA functions through various hierarchies viz an apex body, represented by Secretaries of concerned Ministries/Departments with Secretary, DST being its Chairman. The Apex Body is supported by the Technical Committee, which is a recommending body of NGCMA for GLP certification and technical matters related to GLP. There are trained GLP inspectors who perform GLP inspection in applied labs (NGCMA, GLP n.d.) (Fig. 6.1; Table 6.1).

Terminologies (U.S. FDA Definitions 2018)

Test Facility	:	Includes the personnel, premises and operational units that are involved in non-clinical and safety studies
Test Facility Management (TFM)	:	Person(s) responsible for ensuring that all procedures performed at the test site adhere to the principles of GLP
Sponsor	:	A person or an entity supports a study by providing financial or other resources required for conducting the non-clinical health and environmental safety study
Study Director (SD)	:	The person responsible for the overall conduct of the non-clinical safety studies as per GLP guidelines
Principal Investigator (PI)	:	A person assigned to perform a particular phase of a study, usually in the multi-site study and performs the role of the Study Director at the test site

(continued)

		PI is responsible only for the assigned phase of the study and not for the overall conduct of the study
Quality Assurance Unit	:	Personnel(s)/division responsible to assure that the test facility management that the facility functions in compliance with the GLP principles QA programme remains independent of the study conduction; it is involved in carrying out audits on study, facility and personnel QA is audited by the test facility management
Test System	:	The chemical, biological or physical entity used for evaluating the test item
Test Item	:	An article/substance that is administered/exposed to the test system to obtain data on its safety
Reference Item	:	A known article/substance used for comparison purposes while evaluating the test item safety
Master Schedule	:	It is the test facility's dynamic document which collates, update and review details of the study (GLP/non-GLP), study number, test item category/description, SD/PI details, study initiation and completion date, experimental start and completion date f archiving, study status/remarks and details of individuals involved in a particular study
Study plan	:	The SD signed document that describes the study objectives, experimental design for the conduct of the study and any amendments pertaining to the particular study
Study plan amendment	:	Any intended change(s) made to the study plan after the study has been initiated Study plan amendment should be approved by the SD and circulated to all individuals involved in the study and sponsor
Study plan deviation	:	Any unintended alterations that are raised during the study are called as study plan deviation
Study initiation date	:	The date on which the SD officially signs the study plan before executing the study
Study completion date	:	The date on which the SD signs the final report after completion of the study
Experimental starting date	:	The date on which the first study-specific data is collected
Experimental completion date	:	The date on which the last study-specific data are collected

Table 6.1 showing most common divisions in GLP toxicology lab

S No	Department/divisions	Functions
1	Test Facility Management	Day-to-day administrative activities, to ensure resources (manpower, infrastructure, chemicals and consumables), development of the facility
2	Quality Assurance unit	Preparation of SOPs, audits—study, facility and process-based, to ensure adherence of facility to GLP principles
3	Scientific Unit	Includes animal facility and laboratory scientists viz veterinarians, toxicologists, genetic and eco-toxicologists, biochemists and bioanalytical scientists, veterinary pathologists, microbiologists, etc. involved in various levels of a non-clinical safety study
4	Test Item Control Office	Facility to store the test and reference items equipped with appropriate facilities viz weighing machines, refrigerators and deep freezers, data loggers, dark boxes, fluorescence lights, storage cabins, etc.
5	Archive	Ideally, it is a biometry-accessed facility with fireproof, temperature and humidity-controlled rooms, lockers and cabins with a controlled key system, facility to store documents and materials, biological specimens, and blocks, CD and a hard disc containing information, should have an anteroom for receipt of materials
6	Information Technology and Documentation Control Office	Access controlled facility equipped with validated computers, application software, printers, paper shredders, servers, fireproof and temperature-controlled room.
7	Supportive team	Staff involved in the day-to-day maintenance of heating, ventilation, air conditioning (HVAC), electricity, water lines, emergency (fire, earthquake, flood, etc.) house-keeping and biomedical waste disposal

6.6 Responsibilities of GLP Personnel

1. **Test Facility Management**
   - To plan the scope and the type of studies to be carried out in the facility, and to ensure effective communication with the sponsors
   - To establish the adequate physical plant, instruments and equipment, resources for chemical and consumables, supportive facilities, disaster management plans, third party contracts for testing are available for the facility
   - To ensure implementation of quality assurance program with qualified and experienced personnel and to periodically audit the quality assurance unit/division

- To ensure the appointment of adequate technical manpower (chemistry, toxicology, genetic toxicology, clinical chemistry, pathology) in the facility to carry out the non-clinical safety studies and to issue appropriate job description
- To ensure the appointment of an individual(s) and delegate responsibilities to maintain archive, test item control office, information technology (IT) facilities.
- To maintain floor plan, organogram, master schedule, staff list, instruments and equipment lists and to update as and when required
- To ensure periodical training of staff on their respective field of work and implement staff health monitoring program in the test facility
- To ensure establishment and implementation of technically valid standard operating procedures (SOPs) in the facility
- To assign studies to individuals with required qualifications, training, and experience, and to do so according to a master schedule.
- To designate Principal Investigator with appropriate qualifications, training and experience to supervise the delegated phase(s) in a multi-site study
- To ensure that test and reference items are characterised (Hendriks et al. 2008)

2. **Quality Assurance**
   - To be responsible and ensure that the test facility adheres to the GLP principles
   - To impart periodical training on OECD principles of GLP to test facility staff
   - To maintain copies of organogram, master schedule, floor plan, staff list and ongoing study plans
   - To plan and perform study, facility and process-based audits in the test facility.
   - To perform vendor/supplier validation
   - To report audit findings to concerned staff in the facility and TFM and to ensure the closure of the findings.
   - To review study plan, raw data and study report for compliance with OECD principles of GLP
   - To audit the study report that reflects the raw data
   - To issue a QA statement for inclusion in the final report. (Quality Assurance and GLP | READ online n.d.)
3. **Archivist**
   - To record the receipt of archive documents/materials
   - To ensure that the documents, materials and records archived are systematically indexed, for easy retrieval
   - To ensure ambient environmental facility for document/materials storage. Documents should be routinely checked for dust, moisture, fungus and vermin particles so as to maintain the quality of the archived materials.
   - To establish SOP for retrieval of archived documents/materials during audit/inspection.
   - Before issuing an archived material to any reference reading to the facility staff, the archivist should check the page count of the document, the number of specimens/slides/paraffin blocks, the quantity of test/reference items and size of the electronic data being issued from the archive. When receiving the issued

materials in return, the archivist should verify the documents and ensure it is unaltered.

- The biological specimens stored should be checked for the presence of an adequate quantity of fixatives and ensure the complete immersion of tissues periodically.
- The test/reference item should be checked for any leakage/damage of the container and to perform corrective action if any damage is recorded.
- Electronic data should be checked for readability, dust and scratches periodically.
- Disposal of archived material should be performed by the archivist. It should be packed, labelled properly and collected in appropriate color-coded bags as per the facility SOP (OECD 2007).

4. **Test item control officer**
    - To maintain an entry–exit log, instruments equipment and safety measures in the test item control office
    - To prepare and maintain SOPs relate to the test item control office
    - To ensure safe receipt and storage of the test item as per the conditions and stability specified by the sponsor in the test item data-sheet.
    - To ensure the receipt of the certificate of analysis/test item characterisation data along with the test item.
    - To ensure that the test item receipt note is sent to the sponsor immediately after the test item has been received.
    - In case of the long-term study, a portion of test item should be archived as soon as it is received by the test item control officer
    - To dispense the test item to SD/study personnel for study conduction
    - To track and account for the dispense, wastage and return of test items from the test item control office
    - To ensure all precautionary measures are in place while handling a hazardous substance
    - After completion of the study, the leftover test item should be either returned to the sponsor or disposed-off with the sponsor's consent
5. **Information technology officer**
    - To ensure proper installation of hardware/software, operation and validation of the computer systems.
    - To ensure physical and logical security to computers and IT office
    - To maintain IT-related SOPs, software, installation and validation documents
    - To ensure installation of original application software, including anti-virus program, biostatistics software and update them periodically To create individual login for the user and update password as and when required, that prevent unauthorised access to data and computers
    - To periodically assess the performance of the computer systems as maintenance measures and implementation of the change control process to ensure computer validation status

- To ensure backup procedures for all stored data by using servers and periodical update of external devices (as disaster management measure)
- To review and process the retirement of computer

6. **Study director**
    - To prepare a study plan and to ensure it is signed by the management and sponsor
    - To be responsible for officiating the study with the dated signature of the study plan.
    - To ensure that a copy of the study plan is available to all personnel, including the QA before initiation of the experiment
    - To ensure availability of test item and related documents, test system, animal rooms and necropsy room, study personnel, validated computers, calibrated equipment and instruments, chemicals and consumables, etc, before the start of the study
    - To monitor study personnel adherence to the study plan, study progress, to review raw data and entry of raw data in validated computers
    - To make amendments, if any, and formalise along with immediate information to sponsor
    - To be responsible for the preparation of study report that covering the scientific contents, GLP compliance statement and QA audit summary and statement and, amendments/deviations, if any study, and interpretation of results
    - To archive test item, raw data, study plan and report (hard and soft copy), all tissue specimens, blocks after the completion of the study. (The Role and Responsibilities of the Study Director in GLP Studies I READ online n.d.)
7. **Study Personnel**
    - To adhere to the principles of GLP throughout the study.
    - To assist SD to conduct the study as per the study plan and schedule
    - To assist in dose preparation, dosing, clinical sign assessment, receipt and return of the test item
    - To record raw data promptly, legibly and accurately in formats and to enter in computer systems and report periodically to the SD
    - To document and communicate any deviations in the study plan to SD, and/or if appropriate, the Principal Investigator(s).
    - To exercise health precautions to reduce risk to themselves as well as the study. Any relevant known health or medical conditions that may affect the study should be effectively communicated.
    - To undergo periodical GLP related training (WHO-GLP 2010)
8. **Sponsor**
    - The sponsor shall visit the facility to witness the infrastructure, experience and list of rules (SOPs, formats, etc) of the facility before the sponsoring of the study and monitor the progress during the study.
    - Shall establish a clear line of communication with SD, and TFM during the entire process of the study.

- The sponsor should review the draft study plan and study report and request for the corrections/suggestions, if any with the SD and TFM. The sponsor should sign the final study plan and report.
- In cases where the sponsor provides the test item, the test item should be delivered to the test facility along with a test item data sheet that provides information on the identity, batch no., date of manufacture/expiry, storage, solubility, and certificate of analysis (COA) by the sponsor
- At the end of the life cycle of—study plan/report, test item, the sponsor should provide approval for the receipt or disposal of the same from the test facility (The Role and Responsibility of the Sponsor in the Application of the Principles of GLP I READ online n.d.)

6.7 Test System and Test Item Characterization

Characterisation means establishing the right information about a material used for the experiment. It is necessary to get the maximum possible information about the material used in order to produce reliable and repeatable results. GLP mandates characterisation of test item (typically a chemical molecule) and test system (often a live animal) in non-clinical safety studies.

6.7.1 Test System

The test systems frequently used in GLP studies are microorganisms and animal models. The authenticity of the strain plays a key role in the study outcome.

Microorganism: Purity, phenotypic characterisation, growth characteristics, genotypic characterisation, sensitivity to selected mutagens

Animals: Breed lines, strain confirmation, date of birth, sex, random—biochemistry, serology, pathology, infection

6.7.2 Test Item and Reference Item

In general, the test item refers to pharmaceuticals (human and veterinary drugs), industrial chemicals, agrochemicals (pesticide/herbicide), cosmetic products, food and feed additives, extract from plants, etc. The analytical profile of the test item is most frequently characterised e.g. chemical identity (code, Chemical Abstracts Service Registry Number [CAS number]), name, purity, composition, batch number, solubility, stability, storage conditions, etc.

Particularly, the chemical nature, handling and stability in solution form impact the bioavailability of the test item following the dosing in the test system. The test facility should develop a form (test item data sheet) that requires to fill the below-mentioned requisite with sufficient data evaluation. The following information should be included on a delivery form:

- Manufacturer/Sponsor name, address
- Name of the test item (code/CAS number)
- Chemical composition/structure
- Physical description (state/colour/molecular weight/specific gravity)
- Purity
- Certificate of Analysis (COA) [including spectral and chromatography data]
- Date of manufacture, batch/lot number, date of expiry
- Gross and net weight of the test item supplied to the test facility
- Packing materials, no of packings
- Solubility or suspendability
- Storage conditions
- Date of dispatch of the test item
- Identity of the person responsible for the dispatch
- Name of the transporter and type of carrier.

The test item should be handled when it arrives at the testing facility according to the test item datasheet. It should be entered into the receipt logbook and weighted to verify the data provided by the sponsor. Any deficiencies or problems relating to the receipt of the test items should be attended immediately. The test item should be stored as specified. A small amount of test items should be archived if it is meant for long-term study. The arrival of the test item should be informed to the research director, and any test item-related documents should be provided. SD should cross-check the information provided by the sponsor. SD should perform a dose formulation trial before experimental dosing initiation, to ensure the solubility and homogeneity of the dose formulation. The test item control officer should enter the quantity of test item dispensed to SD or study personnel as and when dispensed and it should be auditable (OECD 2018).

6.8 OECD Documents

The below-mentioned table (Table 6.2) provides information on the series of documents that are published by OECD, which gives information on the testing and assessment of chemicals, guidance on GLP monitoring, inspection, roles and responsibilities of GLP personnel, operational procedures of archives and computer systems in a GLP facility (Fig. 6.2; Table 6.3).

Table 6.2 OECD series of GLP and compliance monitoring documents (OECD GLP n.d.)

Document	Title	Purpose
No. 1	OECD Principles on Good Laboratory Practice	This document provides complete information on the OECD Principles of Good Laboratory Practice
No. 2	Revised Guides for Compliance Monitoring Procedures for Good Laboratory Practice	This document provides information that should be followed while establishing a GLP lab
No. 3	Revised Guidance for the Conduct of Laboratory Inspections and Study Audits	This is a guidance document on the conduction of lab inspections and study audits following the Principles of GLP
No. 4	Quality Assurance and GLP	This document provides information role and responsibility of the quality assurance programs to meet GLP principles
No. 5	Compliance of Laboratory Suppliers with GLP Principles	This document provides information to TFM and suppliers on the basic requirements to meet GLP standards
No. 6	The Application of the GLP Principles to Field Studies	This document provides information for field studies, roles and responsibilities of SD, PI, TFM, QA, etc. to meet the GLP requirements
No. 7	The Application of the GLP Principles to Short Term Studies	This document provides guidance on the conduction and interpretation of data in short term studies
No. 8	The Role and Responsibilities of the Study Director in GLP Studies	Guidance document on the roles and responsibilities of SD
No. 9	Guidance for the Preparation of GLP Inspection Reports	This document provides guidance on elements and/or concepts that helps in the preparation of action taken report following GLP inspection and study audit.
No. 10	The Application of the Principles of GLP to Computerised Systems	This document specifies the requirements for the validation of computerised system in the GLP facility
No. 11	The Role and Responsibility of the Sponsor in the Application of the Principles of GLP	This document defines the role and responsibility of the sponsor in the GLP study
No. 12	Requesting and carrying out inspections and study audits in another country	This document sets out the procedures to be followed during inspections and study audits in a foreign country under the Principles of GLP and compliance monitoring.
No. 13	The Application of the OECD Principles of GLP to the Organisation and Management of Multi-site Studies	This document provides guidance to adhere to GLP principles during multi-site studies
No. 14	The Application of the Principles of GLP to in vitro Studies	This document provides guidance on the interpretation and management of in vitro studies

(continued)

Table 6.2 (continued)

Document	Title	Purpose
No. 15	Establishment and Control of Archives that Operate in Compliance with the Principles of GLP	This document provides guidance for establishing and control of archives facility, use contract archives and guidance for quality assurance units for the GLP compliance on the same
No. 16	Guidance on the GLP Requirements for Peer Review of Histopathology	This document provides guidance for the GLP personnel on the peer review of histopathology and reporting preparation
No. 17	Application of GLP Principles to Computerized Systems	This document currently replaces Document 10 (as of May 2021)
No. 18	OECD Position Paper Regarding the Relationship between the OECD Principles of GLP and ISO/IEC 17025	This document provides information on the possible relationship between the OECD principle of GLP and ISO/IEC 17025
No. 19	Management, Characterisation and Use of Test Items	This document provides guidance on the operation of TICO, characterisation of test and reference items
No. 20	Guidance for Receiving Authorities on the Review of the GLP Status of Non-Clinical Safety Studies	This document provides guidance for assessors on the evaluation of the GLP compliance status of nonclinical safety studies submitted for regulatory purposes.
No. 21	OECD Position Paper Regarding Possible Influence of Sponsors on Conclusions of GLP Studies	This document provides information on the necessity of documentation on the relationship between the TFM and sponsor and advice to generate unbiased data as per GLP requirement

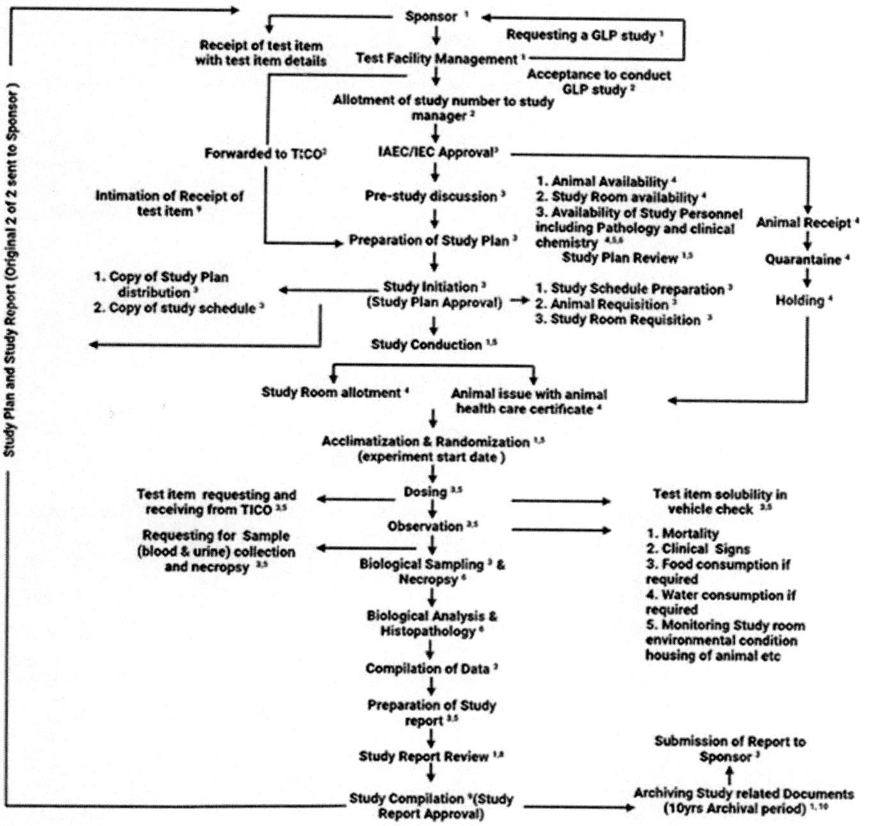

Responsible person: 1-Sponsor,2-Test facility management, 3-Study director, 4-clinical veterinarian, 5-Study personal, 6- Veterinary pathologist, 7- Clinical chemist, 8- Quality assurance unit, 9- Test item officer, 10- Archivist.

Fig. 6.2 Flow chart showing typical procedures followed in conducting a GLP study

Table 6.3 A simple table showing stepwise implementation of GLP in a new facility

Step	Process	Functions and remarks
I.	To fix the purpose and scope of the GLP facility	To establish the GLP standard facility
II.	Appointment of minimum staff as the project team with appropriate experience in TFM, QAU, toxicology, supportive areas	To design facility floor plan, decide instruments and equipment, collection and compilation of documents pertaining to GLP preparation and principles, test guidelines, a framework for SOPs, raw data format, policies, etc.
III.	Completion of building and validation of animal facility, identifying gaps in new labs and other vital units of the test facility and regulatory ethical approval of the animal facility	To review the project status
IV.	Establishment of TICO, archive, biochemistry, pathology, genetic toxicology labs and IT facilities,	To prepare the respective unit as per GLP requirements
V.	Appointment of staff for the new labs and arranging in-house and external GLP training	To prepare the staff for GLP studies
VI.	Purchase of new instruments and equipment and to make the staff familiar to use	To prepare staff to use instruments and equipment in ease
VII.	TFM, QAU and team leads in each section to initiate preparation of the study, process and facility-based SOPs as planned earlier	To train the staff to prepare SOPs and to work as per the SOPs
VIII.	Review and evaluation of SOPs and formats by QAU	To implement the usage of SOPs and formats
IX.	Preparation of organogram and fixing staff with clear job description	To fix staff to GLP adherence
X.	To prepare study plan templates, study formats	To get practiced to using the study plan and raw data formats
XI.	To initiate and complete validation studies	Preparedness to perform studies as per GLP principles and test guidelines
XII.	To initiate GLP studies	To provide full-fledged GLP experience and to do study as a study plan and time frame
XIII.	To have GLP mock drills	To find gaps and to finetune processes as required and to build confidence to face GLP inspection
XIV.	To apply for GLP pre-inspection	For GLP certification
XV.	Addressing non-compliances (NCs), if any and submission of action taken report (ATR)	To proceed for GLP main inspection
XVI.	To apply GLP main inspection	To ensure GLP certification, if no NCs
XVII.	Annual surveillance inspection after 1 year	To update the technical processes
XVIII.	The facility will be audited for GLP compliance once in 3 years	To work on re-certification

References

Baldeshwiler AM (2003) History of FDA good laboratory practices. Qual Assur J 7:157–161. https://doi.org/10.1002/qaj.228

Hendriks R, Van den Eynde H, Coussement W (2008) Responsibilities of test facility management and sponsor in a GLP environment. Ann Ist Super Sanita 44:407–408

National GLP, Compliance Monitoring Authority (NGCMA), Department of Science & Technology (n.d.). https://dst.gov.in/about-ngcma

OECD (2007) Establishment and control of archives that operate in compliance with the principles of GLP, OECD Series on Principles of Good Laboratory Practice and Compliance Monitoring. OECD. https://doi.org/10.1787/9789264084995-en

OECD (2018) Test and reference/control items. In: Guidance Document on Good In Vitro Method Practices (GIVIMP), OECD Series on Testing and Assessment. OECD, pp 97–120. https://doi.org/10.1787/9789264304796-11-en

OECD GLP (n.d.). https://www.oecd.org/chemicalsafety/testing/overview-of-good-laboratory-practice.htm. Accessed 19 July 2021

Quality Assurance and GLP | READ online (n.d.). oecd-ilibrary.org. https://read.oecd-ilibrary.org/environment/quality-assurance-and-glp_9789264078598-en. Accessed 31 July 2021

The Role and Responsibilities of the Study Director in GLP Studies | READ online (n.d.). oecd-ilibrary.org. https://read.oecd-ilibrary.org/environment/the-role-and-responsibilities-of-the-study-director-in-glp-studies_9789264078673-en. Accessed 31 July 2021

The Role and Responsibility of the Sponsor in the Application of the Principles of GLP | READ online (n.d.). oecd-ilibrary.org. https://read.oecd-ilibrary.org/environment/the-role-and-responsibility-of-the-sponsor-in-the-application-of-the-principles-of-glp_9789264078734-en. Accessed 31 July 2021

U.S. FDA Definitions (2018). FDA

WHO-GLP (2010) GLP trainer. World Health Organization, Geneva